Marine Mammals: A Comprehensive Study

Timothy Garner

Larsen & Keller
www.larsen-keller.com

Marine Mammals: A Comprehensive Study
Timothy Garner
ISBN: 978-1-64172-617-7 (Hardback)

目 Larsen & Keller

Published by Larsen and Keller Education,
5 Penn Plaza,
19th Floor,
New York, NY 10001, USA

Cataloging-in-Publication Data

Marine mammals : a comprehensive study / Timothy Garner.
    p. cm.
Includes bibliographical references and index.
ISBN 978-1-64172-617-7
1. Marine mammals. 2. Aquatic mammals. 3. Marine animals. I. Garner, Timothy.
QL713.2 .M37 2022
599.5--dc23

For more information regarding Larsen and Keller Education and its products, please visit the publisher's website www.larsen-keller.com

Table of Contents

Preface

This book is a culmination of my many years of practice in this field. I attribute the success of this book to my support group. I would like to thank my parents who have showered me with unconditional love and support and my peers and professors for their constant guidance.

Marine mammals are the aquatic organisms that depend on the ocean and other marine ecosystems for their survival. Such organisms include sea otters, seals, manatees, whales and polar bears. The adaptation of marine mammals to an aquatic lifestyle varies from species to species. Some mammals such as cetaceans and sirenians are fully aquatic. Animals such as seals and sea lions are semiaquatic. They spend a major part of their life inside water but they need to come to land to perform some specific activities such as molting, breeding and mating. Some marine mammals such as polar bears and otters are much less adapted to an aquatic lifestyle. Marine mammals have numerous anatomical and physiological features which assist them in overcoming the challenges associated with aquatic living. The topics included in this textbook on marine mammals are of utmost significance and bound to provide incredible insights to readers. Also included herein is a detailed explanation of the various concepts and theories related to this subject. This book is an essential guide for both academicians and those who wish to pursue this discipline further.

The details of chapters are provided below for a progressive learning:

Chapter – Introduction

The aquatic mammals that depend on the ocean and other marine ecosystems for their existence and survival are known as marine mammals. Some of the various types of marine mammals belong to cetacea order and sirenia order. This is an introductory chapter which will introduce briefly all the significant aspects of these types of marine mammals.

Chapter – The Evolution of Marine Mammals

The marine mammals are broadly categorized into cetaceans, pinnipeds and sirenians. These marine mammals have evolved from different land mammals. The topics elaborated in this chapter will help in gaining a better perspective about the evolution of these marine mammals.

Chapter – Cetaceans

Cetaceans are warm blooded aquatic mammals who have well-developed senses and a layer of fat under skin to maintain body heat in cold water. Some of the common cetaceans are whales, porpoises and dolphins. This chapter has been carefully written to provide an easy understanding of these cetaceans as well as their ecology and physiology.

Chapter – Pinnipeds

Pinnipeds are a diverse category of carnivorous, fin-footed, semiaquatic marine mammals. They have streamlined bodies and four limbs which are evolved into flippers. Some of the different types of pinnipeds are sea lions, eared seals and walruses. The chapter closely examines the key features of these types of pinnipeds to provide an extensive understanding of the subject.

Chapter – Sirenians

Sirenians are an order of fully aquatic, herbivores mammals that live in rivers, swamps, estuaries and marine wetlands. Some of the different types of sirenians are West Indian manatees, African manatees and dugongs. This chapter discusses in detail these types of sirenians and their characteristics.

Chapter – Otters and Polar Bears

Otters are the carnivorous mammals that are semiaquatic, aquatic or marine. Polar bears are hyper carnivorous bears that generally inhabit the Arctic Circle. They have a well-developed sense of smell as well as good vision. The topics elaborated in this chapter will help in gaining a better perspective about the various characteristics of otters and polar bears.

Chapter – Threats to Marine Mammals and their Protection

There are numerous human activities which threaten marine mammals. A few of these are ship strikes, acoustic pollution, open net fishing, commercial hunting, oil spills, etc. Some of the steps taken to protect them are the Marine Mammal Protection Act and the establishment of marine protected areas. All these diverse threats to marine mammals and the measures to protect them have been carefully analyzed in this chapter.

Timothy Garner

Introduction

The aquatic mammals that depend on the ocean and other marine ecosystems for their existence and survival are known as marine mammals. Some of the various types of marine mammals belong to cetacea order and sirenia order. This is an introductory chapter which will introduce briefly all the significant aspects of these types of marine mammals.

MARINE MAMMALS

Marine mammals evolved from their land dwelling ancestors over time by developing adaptations to life in the water. To aid swimming, the body has become streamlined and the number of body projections has been reduced. The ears have shrunk to small holes in size and shape. Mammary glands and sex organs are not part of the external physiology, and posterior (hind) limbs are no longer present.

Mechanisms to prevent heat loss have also been developed. The cylindrical body shape with small appendages reduces the surface area to volume ratio of the body, which reduces heat loss. Marine mammals also have a counter current heat exchange mechanism created by convergent evolution where the heat from the arteries is transferred to the veins as they pass each other before getting to extremities, thus reducing heat loss. Some marine mammals also have a thick layer of fur with a water repellent undercoat and/or a thick layer of blubber that can't be compressed. The blubber provides insulation, a food reserve, and aids with buoyancy. These heat loss adaptations can also lead to overheating for animals that spend time out of the water. To prevent overheating, seals or sea lions will swim close to the surface with their front flippers waving in the air. They also flick sand onto themselves to keep the sun from directly hitting their skin. Blood vessels can also be expanded to act as a sort of radiator.

One of the major behavioral adaptations of marine mammals is their ability to swim and dive. Pinnipeds swim by paddling their flippers while sirenians and cetaceans move their tails or flukes up and down.

Some marine mammals can swim at relatively high speeds. Sea lions swim up to 35 kph and orcas can reach 50 kph. The fastest marine mammal, however, is the common dolphin, which reaches speeds up to 64 kph. While swimming, these animals take very quick breaths. For example, fin whales can empty and refill their huge lungs in less than 2 seconds. During dives marine mammals' larynx and esophagus close automatically when they open their mouths to catch prey. Oxygen is stored in hemoglobin in the blood and in myoglobin in the muscles. The lungs are also collapsible so that air is pushed into the windpipe preventing excess nitrogen from being absorbed into the tissues. Decreasing pressure can cause excess nitrogen to expand in the tissues as animals ascend to shallower depths, which can lead to decompression sickness aka "the bends." Bradycardia, the reduction of heart rate by 10 to 20%, also takes place to aid with slowing respiration during dives and the blood flow to non-essential body parts. These adaptations allow sea otters to stay submerged for 4 to 5 minutes and dive to depths up to 55 m. Pinnipeds can often stay down for 30 minutes and reach average depths of 150-250 m. One marine mammal with exceptional diving skills is the Weddell seal, which can stay submerged for at least 73 minutes at a time at depths up to 600 m. The length and depth of whale dives depends on the species. Baleen whales feed on plankton near the surface of the water and have no need to dive deeply so they are rarely seen diving deeper than 100 m. Toothed whales seek larger prey at deeper depths and some can stay down for hours at depths of up to 2,250 m.

Marine mammals are often very social animals. Dolphins travel in pods (schools) and catch rides on the bow waves of boats. Marine mammals are also known to help each other when one member of the group is injured. There have been accounts of members of a pod refusing to leave the wounded or dying, a trait often exploited by whalers. Cetaceans (whales and dolphins) often hunt together, often with one leading the pod to act as a scout when entering unfamiliar territory. This close knit socialization is thought to be a factor in some whale strandings when a pod follows one or more members of the group that have become disoriented due to storm, illness, or injury.

Many marine mammals also participate in yearly migrations, either in groups or individually. Toothed whales are an exception and only move about in search of food, but some baleen whales (such as gray whales) embark on extremely long migrations, moving from tropical breeding grounds in winter to feeding areas in colder waters during the summer.

Communication

Marine mammals are capable of sophisticated communication because they live in a world dominated by sound, which travels much more efficiently through water than through air. Dolphins communicate with sound to coordinate hunts; humpback whales sing to attract females. Female pinnipeds and their pups recognize each other by their "voices." Slapping the surface during breaching can be heard for miles. Whales have no vocal cords; they warble for up to 30 minutes between breaths just by recycling air. They also emit low frequency sounds that can be heard by humans such as grunts, barks, squeaks, chirps, or even moos. These noises are thought to be associated with different moods and are believed to be used as social or sexual cues during communication. They might also serve as a signature to allow one animal to be recognized by another. Certain pods are known to even have dialects than can be distinguished from others.

Echolocation is a skill that only toothed cetaceans, bats and a few birds have perfected. They send out rapid sound pulses and listen to their echo to find prey and determine their surroundings. It is thought that sperm whales also use echolocation to stun squid with loud clicks. Clicks can be repeated at different frequencies with low frequencies traveling long distances that are highly penetrating. Toothed whales have a structure called the melon on their forehead that focuses and directs the sound waves; incoming sounds are received primarily in the lower jaw, which is filled with fat or oil that transmits the sound to the inner ear.

Evolution

About 65 million years ago (mya) when dinosaurs became mostly extinct, marine mammals began to evolve from their land-dwelling ancestors. Their evolution into sea-dwelling mammals is thought to be a result of the availability of new marine food sources and a way to escape from their terrestrial predators. The fossil record for whales is not as extensive as it is for other marine mammals such as otters and pinnipeds, therefore the transition period between land and water is unclear. In 1994, the remains of Ambulocetus natans ("the walking whale that swam") dating 49 mya were found in Pakistan in what's left of the Tethys Sea. These whale remains showed that the animal once had strong legs with long feet, similar to modern pinnipeds, that were functional both on land and in the sea. It retained a tail, but lacked flukes, however it is still thought that this animal swam like modern whales by moving the rear portion of its body up and down. In 2001, other fossils were found that linked early cetaceans to hoofed animals (ungulates).

Life History

All mammals are viviparous, meaning that their eggs develop inside the female and the embryo derives nutrition from the mother. Whales and pinnipeds usually mate and give birth in the spring, with pregnancies lasting between 12-18 months. Seals usually have a single pup each year. Whales, however reproduce more slowly and generally raise one calf every 1-3 years. Cetacean calves are born tail first to keep them attached to the placenta as long as possible to avoid oxygen deprivation.

Mating can be a social as well as functional activity with marine mammals. With dolphins, sex is used to establish and maintain bonds among the group. Humpback and beluga whales both take part in group matings.

CHARACTERISTICS OF MARINE MAMMALS

The life of marine mammals is very different from that of terrestrial mammals. In order to survive in this environment, these animals have acquired special characteristics during their evolution.

Water is a much denser medium than air and, in addition, it offers greater resistance. This means that aquatic mammals have developed extremely aerodynamic bodies that allow them to unwind easily. Developing fins similar to those of fish has been a predominantly significant morphological change. It allows these animals to: increase speed, direct swimming and communicate.

Water is a medium that absorbs much more heat than air, therefore, aquatic mammals have a thick layer of fat under their hard and robust skin. Also, it serves as warmth and protection when they live in specifically cold areas. Some marine mammals come into more contact with air, because certain vital functions take them out of the water, such as reproduction.

Those marine mammals that, at certain periods of their lives, have inhabited the great depths, have developed other organs which give them the ability to be able to live in darkness. Such abilities include incredible senses; such as, sonar and echolocation. Some senses, however, such as sight are not useful in these ecosystems. This is because sunlight does not reach such depth.

Like all mammals, these aquatic animals have: mammary glands that produce milk for their offspring, sweat glands and they gestate their offspring inside of their bodies.

How do Marine Mammals Breathe?

Aquatic mammals need air to breathe. Therefore, they take large amounts of air and keep it inside of their lungs for long periods of time. Once they submerge after taking air, they are able to redirect blood to the brain, heart and their skeletal muscle. Their muscles hold a high concentration of a protein called myoglobin, which can accumulate large amounts of oxygen.

In this way, aquatic animals are able to remain underwater for considerable periods of time without breathing in air. Young and newborn marine mammals take a while to developed this capacity, therefore, they need to breaths more often than developed mammals.

Types of Aquatic Mammals

Most species of aquatic mammals live in the marine environment. There are three orders of aquatic mammals, which include: cetacea, carnivora and sirenia.

- Cetacea order

  The most representative species within the cetacea order are: whales, dolphins, sperm whales, killer whales and porpoises. The cetaceans evolved from a species of ungulate terrestrial carnivores more than 50 million years ago. The cetacea order is divided into three suborders (one of them extinct):

  ○ Archaeoceti: Terrestrial quadruped animals: predecessors of current (extinct) cetaceans.

  ○ Mysticeti: The baleen whale. They are carnivorous animals without teeth that take large puffs of water. They then filter it through their baleens (filter-feeder system) and catch the trapped fish with their tongues.

  ○ Odontoceti: This includes dolphins, killer whales, porpoises, sperm whales and beaked whales. This group is very diverse, its main charcateristic being that they all have teeth. In this group we also find the Amazon river dolphin (Inia geoffrensis), a species of aquatic river mammal.

- Carnivorous order

  In the carnivora order, we include: seals, sea lions and walruses, sea otters and polar bears. This group of animals appeared about 15 million years ago and it is believed that they are closely related to mustelids and ursidae (bears).

- Sirenia order

  The last order, sirenia, includes: dugongs and manatees. These animals have evolved from the tethytheria clade, animals very similar to the elephants that appeared about 66 million years ago. Dugongs inhabit Australia, the Red Sea and Indian Ocean and, manatees, Africa and America.

Importance of Sound to Marine Animals

In addition to vision, marine animals use other mechanisms,
such as sound, to gather information and communicate.

Hearing is the universal alerting sense in all vertebrates. Sound is so important because animals are able to hear events all around them, no matter where their attention is focused. Many species of blind amphibians, reptiles, fishes and mammals are known, but no naturally profoundly deaf vertebrate species have been discovered. Although hearing is important to all animals, the special qualities of the undersea world emphasize the use of sound.

Sound travels far greater distances than light under water. Light travels only a few hundred meters in the ocean before it is absorbed or scattered. Even where light is available, it is more difficult to see as far under water as in air, limiting vision in the marine environment It is similar to looking through fog on land. So, the best opportunity for long-range vision underwater — especially in murky water — is to swim beneath objects and see their silhouettes.

Touch is very important to marine mammals, such as these manatees.

The undersea world presents very different conditions for hearing as well as seeing. Sound travels much farther underwater than in air. The sounds produced by many marine mammals can project for miles. Strong echoes are always present underwater, because sound travels without much loss and there are many underwater surfaces that reflect sound. So, it can be tricky to communicate using sound underwater, because a listener may have to sort through many different sounds and confusing echoes to hear the message. Marine mammal sounds are probably structured so that they can be recognized in spite of all the echoes.

Of the five senses (touch, taste, smell, sight, hearing), touch probably functions in the most similar way for animals on land and underwater. Touch is very important in close social interactions. For underwater animals, the sense of touch may also provide important information about water currents and the motions of nearby animals.

The senses of taste and smell enable animals to detect chemical compounds; they serve very similar functions. The underwater environment presents very different opportunities for these senses. Chemical particles travel much more slowly in water than in air. Thus, except for very short range effects, it is likely that the sense of smell is largely used to detect the trail of chemicals left behind by a moving animal (or the "trail" created by stationary animal in a current), rather than the chemicals that drift away from a stationary animal.

Underwater sound allows marine animals to gather information and communicate at great distances and from all directions. The speed of sound determines the delay between when a sound is made and when it is heard. The speed of underwater sound is five times faster than sounds traveling in air. Sound travels much further underwater than in air. Thus marine animals can perceive sound coming from much further distances than terrestrial animals. Because the sound travels faster, they also receive the sounds after much shorter delays (for the same distance). It is no surprise that marine mammals have evolved many different uses for sounds.

Marine animals rely on sound to acoustically sense their surroundings, communicate, locate food, and protect themselves underwater. Marine mammals, such as whales, use sound to identify objects such as food, obstacles, and other whales. By emitting clicks, or short pulses of sound, marine mammals can listen for echoes and detect prey items, or navigate around objects. This animal sense functions just like the sonar systems on navy ships. It is clear that producing and hearing sound is vital to marine mammal survival.

Sound is important to fish, such as this rock hind,
for mating, feeding and survival.

Sound is also important to fishes. They produce various sounds, including grunts, croaks, clicks, and snaps, that are used to attract mates as well as ward off predators. For the oyster toadfish, sound production is very important in courtship rituals. Sound is produced by the male toadfish to attract the female for mating and is especially important in the murky waters that toadfish inhabit where sight is limited. Marine invertebrates also rely on sound for mating and protection. Little research has been done on marine invertebrates that produce sounds, but for those that do, like shrimp and lobsters, sound is very important for survival against predators.

As you can see, sound is very important to its underwater inhabitants. Most marine animals rely on sound for survival and depend on their unique adaptations that enable them to communicate, locate food, and protect themselves underwater.

Intraspecific Behavior of Marine Mammals

The heterogeneous phenomenon considered as intraspecific aggressive or agonistic behavior represents a conglomerate of social responses, including male disputes over territorial boundaries, female fights to protect an offspring, female harassment and forced copulations, and infant abuse and killing. Agonistic encounters:

1. Mediate competition for limited resources economically defendable and valuable to the fitness of an individual. Finite resources that can be monopolized would lead to social conflict between individuals of different sexes and generations and of the same sex and similar age class and status. Most often, agonistic confrontation (at least the most conspicuous interactions) involves sexually mature males.

2. Are more common in some social contexts, such as breeding on land in a polygynous mating systems, in which competition for resources is typically solved via aggressive disputes. Size and strength (but also agility) correlate positively with winning a contest through exerting dominance over individuals subdued by the costs of rebellion.

3. Have a broad range of costs for actors and recipients, from simple rejection after a ritualized threat display to injury or even death after an overt physical encounter.

The form and frequency of agonistic behavior partially reflect the sophistication of a social system. Aquatic mammals vary widely in the complexity of their societies, thus in the manifestation of agonistic behaviors. The most openly competitive societies characterize the otariids, the walrus (Odobenus rosmanis), and pho-cids that live in crowded conditions (e.g., elephant seals, Mirounga spp., and gray seals. Halichoerus grypiis), a fertile ground for aggressive social interactions.

Conversely, polar bears (Ursus maritimus), all the mysticetes and river dolphins, and some other phocids (e.g., Ross and leopard seals, Om-niatophoca rosii and Hydrurga leptonyx) generally occur in smaller social groups, except for periods during reproduction in which breeding males engage in scramble competition over receptive females. The most complex social systems in the aquatic mammals would characterize some of the odontocete cetaceans, such as killer whales, Orcinus orca, pilot whales, Globicephala spp., bottlenose dolphins, Tursiops spp., or sperm whales, Physeter niacrocephalus. These species live in stable social units and show coordinated, cooperative behaviors. The long-term shared history among individuals of the group would have ritualized many of the overt aggressive responses typical of the polygynus pinnipeds.

Male-male Competition for Mates

Competition over limited resources to maximize reproductive success would be the most common origin of agonistic encounters. It is likely that in all aquatic mammals, males compete for access to reproductive females, by either direct or indirect monopolization, through achieving the best place for reproduction or the highest status in a dominance rank. Defense systems can set the stage for the evolution of sexually selected traits, such as dimorphism in size and in special morphological structures (e.g., tusks, manes, elongated snouts).

The behavioral manifestations of conflict directed to the intimidation of rivals is often referred to as agonistic display or agonistic social signaling. Behavioral displays include vocal signals, facial expressions, and stereotyped postures and movements. Such as static open-mouth threats, open-mouth sparring. foreflipper raise or waving, and oblique staring. Overt fighting is commonly a last-resort solution to conflict.

Pinnipeds

The form of male agonistic encounters and their outcome lias been described in detail for several pinnipeds. Within the highly polygynous otariids and phocids there are examples of resource-defense (territorial) and female-defense polygynous systems. Both types of polygyny may occur in the same species, such as in the South American sea lion, Otaria flavescens, as a function of different ecological conditions.

The establishment and defense of a territorv involve vocal displays, stereotyped postures and movements, and fights. During territorial displays, male contenders may rush toward each other with the mouth open or vocalizing, weave the head from side to side, puff out the chest, or perform the "oblique stare" posture at one another, but physical contact is usually avoided. Much of the fighting between otariid males takes place early in the breeding season, when territorial boundaries are being established. When physical contact occurs, it typically lasts a lew seconds but may be violent, particularly in the largest sea lions. Fights involve chest-to-chest pushing, vigorous biting of the neck and face, lunging, and slashing at the opponent's flippers, chest, and hindquarters.

In female-defense polygyny, females cannot be sequestered or attracted to a particular place. Males then compete to achieve a position among the females in the breeding colony and move with the shifting population of females. Association to a particular group of females is loose and may change even during the same day due to female redistribution related to the physical environment (high temperatures, variable space due to tidal movements) or to social behaviors (e.g., group raids of ousted males into the colony).

In phocids such as elephant seals, males aggressively establish a dominance hierarchy rather than a resource or female defense system. Only the highest ranking individuals have undisturbed reproductive access to females. During the establishment of hierarchies, males attempt to intimidate each other with vocal displays. If none of the contestants retreats, then a chest-to-chest fight takes place. Fights in elephant seals are violent confrontations and may last half an hour. Each bull throws his weight against the other and slashes at his opponent's face, neck, and back with long canines. Most fights end with multiple lacerations and bloody wounds or even a broken canine tooth; even death may occur on rare occasions.

Vocal threats are a common component of agonistic encounters. Pinnipeds vocalize both in air and underwater. Harp (Pagophilus groenlandieus), ringed (Ptisa hispida). Weddell (Leptonijchotes weddellii), and bearded (Eiig-nathus barbatus) seals and the walrus have a rich underwater vocal repertoire. Males maintain underwater territories and vocalizations seem to be part of territorial displays. Vocal displays can be of a repetitive nature and are then termed songs. Otariid males, particularly among the fur seals (Arctocephalus spp.), have a rich variety of airborne threat vocalizations associated with boundary display postures. The California sea lion, Zalo-phns californianus, vocalizes both in air and underwater (several phocids also produce airborne and underwater sounds).

The strong airborne calls or barks of Zalophus occur during breeding and nonbreeding seasons and may serve to advertise dominance. In elephant seals, airborne threat displays consist of loud and directional pulsed sounds that tend to precede fights.

Cetaceans

There is comparatively little description of agonistic encounters in the rest of the aquatic mammals. Agonistic behaviors to establish dominance relationships were described among dolphins in captivity. Observations of free-ranging cetaceans described a range of behaviors interpreted as agonistic, such as lobtailing, tail and flipper slaps to the body of other individuals, open-mouth postures, jaw claps, forceful exhalations, chases, body charges and leaps and body slaps, and vocal threat displays. Escalated agonistic displays involve striking with flukes, biting, and jousting with tusks, the latter in narwhals (Monodon monoceros). Humpback whale (Megaptcra novneangliae) males fight vigorously in surface-active groups and receive not only scrapes and scratches but also deep gouges and bloodv wounds as a result.

The scar pattern of some odontocetes has been interpreted as the consequence of tooth marks and violent interactions. Several odontocetes have conspicuous scars. In Risso's dolphins, Grampus griseus, narwhals and several of the beaked whales, most of the body is covered with scars. At least for the narwhal, scars have been associated with intraspecific agonistic encounters. Scrape marks are also common in baleen whales. It has been suggested for the southern right whales, Eubnlaena australis, that males may use the thorny callosities during scramble competition over females.

Agonistic contests in cetaceans also involve vocal displays. Males of the humpback whale escort receptive females and vigorously rebuff other males bv threatening displays such as thrashing of the flukes. The underwater songs of humpback whales in Hawaiian breeding grounds are performed by males and likely serve as communication signals in the context of male competition.

An example of male-male competition involves Australian bottlenose dolphins, Tursiops aduncus. Males of this population form stable alliances of two or more individuals that cooperate to obtain and control reproductive females. Two alliances occasionally combine efforts to sequester or defend females from another alliance. Alliances in dolphins and group raids in sea lions represent special cases in which competition involves the participation of several individuals simultaneously.

Other Aquatic Mammals

Sea otters (Enhtjdra lutris), polar bears, and sirenians tend to be more solitary or live in low-density societies with little interaction among individuals. Male sea otters are polygynous, establish breeding territories, and mate in the water. Females live in low-density areas chosen in relation to the distribution and abundance of food.

During the mating season, polar bear males rove to locate receptive females that are dispersed and solitary. Males access one female at a time. Competition involving physical interactions has been observed rarely but is indicated by broken teeth and scarring on the head and neck.

Manatees (Trichechus spp.) form mating groups in which several males compete for access to a receptive female by pushing and shoving each other. Physical competition for females also occurs in dugongs (Dugong dugon) with some males obtaining scars probably made by the tusks of other males.

Tusks as Special Structures for Aggression

Two species of marine mammals have extraordinarily developed tusks: the walrus and the narwhal. The two upper canines in both male and female walrus are extraordinarily elongated. The massive tusks of a male can weigh up to 10 pounds and be almost 1 m long. Both sexes use tusks in squabbles to threaten one another, and. perhaps, to establish dominance. Males may force their way to selected places in crowded colonies by pushing and jabbing other walruses with their tusks.

The tusks of narwhals are even more exceptional morphological traits. As a general rule, the left canine in males extends anteriorly into a spiraled tusk to a length that may exceed 2.5 m. Some males have two tusks and a few females also develop one or even two shorter and less robust tusks. It has been suggested that narwhal tusks may be used to disturb or pierce prey, to open breathing holes in the ice, as defense weapons against predators, or as organs of sexual display. Although tusks may be used in more than one context, evidence shows that they serve in aggressive encounters. Evidence includes direct observations of males crossing tusks and striking them against one another, scar patterns (with adult males having more and larger scars on the head after attaining sexual maturity), significantly higher incidences of broken tusks in mature males compared to immature individuals or females, and imbedded splinters and tusk tips found in the head of males. Tusks are also used to spear individuals of other species or, apparently, at times even female narwhals.

Sexual Selection and Special Morphological Traits

Pronounced sexual dimorphism in the direction of males being heavier and larger than females is common in all otariids, the walrus, and some phocids (e.g., elephant and gray seals). This kind of dimorphism often indicates direct physical confrontation among reproductive males involving pushing or strength contests. Dimorphism is not apparent, is slight, or is even reversed in most other phocids. A lack of or even reversed dimorphism is often accompanied by the defense of aquatic territories, aquatic mating, and serial monogamy. Females in these species are usually dispersed and breeding occurs over a protracted period. Social and ecological conditions do not favor frequent direct physical confrontations, but competition does occur, and may for more agile rather than larger individuals.

Among other aquatic mammals, males are much larger than females in some odontocetes, such as killer and sperm whales, whereas dimorphism is reversed in all the mysticetes. Mysticetes may have promiscuous mating systems in which competition for insemination takes place at the level of males displacing each other from the vicinity of a female and of spenn cells displacing or diluting sperm of other males. Gray (Eschrichtius robustus), right (Eubalaena spp.), and bowhead (Balaena mysticetus) whales have larger testes than expected based on their body weight, suggesting selection for sperm competition.

In addition to dimorphism in body size, males of some species evolved special secondary sexual features that may function in the context of competition for mates. Examples include the enlarged snouts of male elephant seals and gray seals and the inflatable nasal cavity of hooded seals (Cystophora cristata). Hooded seal males can blow a red, balloon-like sac from one nostril that is similar in shape to the long proboscis of elephant seals. These organs have visual or acoustic effects and may allow other males and females to judge the quality of a contender or a sexual partner. The developed neck and mane of sea lions with long and thick guard hairs also has visual effects and serves as a shield that protects internal organs from bites.

Avoiding Fights

Competition for resources by direct aggression is a costly experience in species capable of inflicting serious injuries that could lower future fitness of the contestants. Thus, contenders with low chances of success should avoid physical confrontations. Theory predicts that the assessment of the fighting ability of competitors and of resource value prior to an escalation of violence may allow differential adaptive responses on the basis of the perceived asymmetries. Once a territory or social hierarchy is established, disputes tend to be asymmetric contests in which territory owners or high-ranking males almost always win. Threat displays may then serve as indicators of a quality and motivational state of a contender. Individual variation in vocal displays may help territorial males to recognize one another and to forgo direct competition if each knows its respective status.

In female-defense systems, the proportion of sexually receptive females accessible to a male is variable in space and time. Thus, the level of asymmetry can vary within the same day of a breeding season. This social context would favor behaviors that are unusual in strict territorial or hierarchical systems, such as group raids in South American sea lions.

Group Raids and other Forms of Male Harassment of Reproductive Females

In the South American sea lion, losers in male-male competitions at times raid breeding colonies in groups of dozens of individuals. Raiders abduct females from the harems of established males and attempt to mate with them. A male seizes a female in his jaws and hurls her into the air to a spot where he can hold his ground against other males while aggressively keeping her in place. In the process, females are often wounded and can be killed. Perhaps group raids represent a primitive stage of a male alliance or coalition.

Violent behavior toward females is relatively common in pinnipeds. Harassed females are injured and sometimes killed by males during mating attempts. Le Boeuf and Mesnick suggested some social conditions that can increase mortality risks to a female during mating: (a) marked male sexual dimorphism, (b) males outnumbering females, (c) use of force or potentially dangerous weapons in mating, and (d) monopolization of mating by a few individuals through direct or indirect control of resources (space, females, food, etc.) with forcible exclusion of the majority of the competitors. All of these traits are common in the most polygynous mating systems.

The majority of female deaths during the breeding season of elephant seals, the most sexually dimorphic of all the pinnipeds, occurs by traumatic injuries inflicted by males during mating attempts as the females depart the harems for the sea at the end of lactation. Male South American sea lions and elephant seals are three to five times heavier than females, have large canines, and often bite the neck of the female when copulating. Breeding colonies early and late in the season have a high number of males that intercept departing females and attempt to mate with them. Mating injuries inflicted by males to females have also been reported for several other species. Male aggression toward females may be a selective force in shaping female behavior, female choice, maternal performance, and reproductive synchrony.

Female Agonistic Behavior

In polygynous pinnipeds, females are aggressive toward one another and rarely tolerate neighbors close by, which helps to regulate density of a site. A common context of female agonistic encounters is that of protection of a pup in a crowded breeding colony Alien pups are often bitten by females. Aggressive mothers react rapidly and intensively to the threat to their pup by a neighbor, which enhances chances of pup survival by decreasing the risks of mother-pup separation and pup injury.

At times, females threaten transient males when the latter approach or protest vocally when males mount them. As a result, a harassing male will then be more likely challenged by another male who hears the female vocalizing. These challenges generally interrupt a male's approach or mount, and hence a potential copulation. By resisting male copulatory attempts, females increase their likelihood of mating with a dominant individual, which may be viewed as an indirect form of mate choice.

Abuse and Killing of Young

Infanticide is the killing by conspecifics of young still dependent on their mothers. Infant abuse implies injury of a young either via active violent behaviors or via passive neglect. Violent abuse of pups by males (most often young individuals but also adults) occurs in several pinniped species, particularly in sea lions and elephant seals. The killing of young is most often the by-product of abuse, although it may also occur as a directed behavior. In addition to pinnipeds, infanticide has been described in polar bears and is inferred in at least one odonto-cete, the common bottlenose dolphin.

References

- Marine-mammals, marine-vertebrates, creatures: marinebio.org, Retrieved 2 March, 2019
- Aquatic-mammals-list-characteristics-and-examples-2893: animalwised.com, Retrieved 3 April, 2019
- Why-is-sound-important, importance-of-sound, animals: dosits.org, Retrieved 4 May, 2019
- Aggressive-behavior-intraspecific-marine-mammals, marine-mammals, Retrieved 5 June, 2019

The Evolution of Marine Mammals

The marine mammals are broadly categorized into cetaceans, pinnipeds and sirenians. These marine mammals have evolved from different land mammals. The topics elaborated in this chapter will help in gaining a better perspective about the evolution of these marine mammals.

Although all marine mammals evolved from land mammals, it may surprise some that each group of marine mammals has its own unique ancestry. There are three groups of marine mammals: the cetaceans (whales, dolphins and porpoises), the pinnipeds (fur seals, sea lions, walruses and seals) and the sirenians (dugongs and manatees). Through convergent evolution, each of these groups separately evolved similar body structures as they adapted to a life in the marine environment.

EVOLUTION OF PINNIPEDS

The name pinniped refers to the paddle-like fore- and hind limbs of seals, sea lions, and walruses, which they use in locomotion on land and in the water. Pinnipeds spend considerable amounts of time both in the water and on land or ice, differing from cetaceans and sirenians, which are entirely aquatic. In addition to blubber, some pinnipeds have a thick covering of fur.

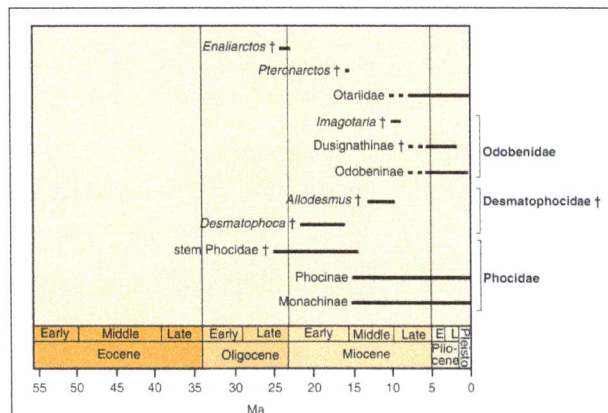

Chronological ranges of extinct and
living pinnipeds. Ma = million years ago.

In seeking the origin of pinnipeds we must first define them. Is the group monophyletic or not? Although this question has been subject to considerable controversy during the last century, the majority of scientists today agree that the Pinnipedia represent a natural, monophyletic group. Pinnipeds are diagnosed by a suite of derived morphological characters. All pinnipeds, including both fossil and recent taxa, possess the characters described later, although some of these characters have been modified or lost secondarily in later diverging taxa. Some of the well known synapomorphies possessed by are defined as follows:

- Large infraorbital foramen: The infraorbital foramen, as the name indicates, is located below the eye orbit and allows passage of blood vessels and nerves. It is large in pinnipeds in contrast to its small size in most terrestrial carnivores.

- Maxilla makes a significant contribution to the orbital wall. Pinnipeds display a unique condition among carnivores in which the maxilla (upper jaw) forms part of the lateral and anterior walls of the orbit of the eye. In terrestrial carnivores, the maxilla is usually limited in its posterior extent by contact of several facial bones (jugal, palatine, and/or lacrimal).

- Lacrimal absent or fusing early in ontogeny and does not contact the jugal. Associated with the pinniped configuration of the maxilla is the great reduction or absence of one of the facial bones, the lacrimal. Terrestrial carnivores have a lacrimal that contacts the jugal or is separated from it by a thin sliver of the maxilla and thus can be distinguished from pinnipeds.

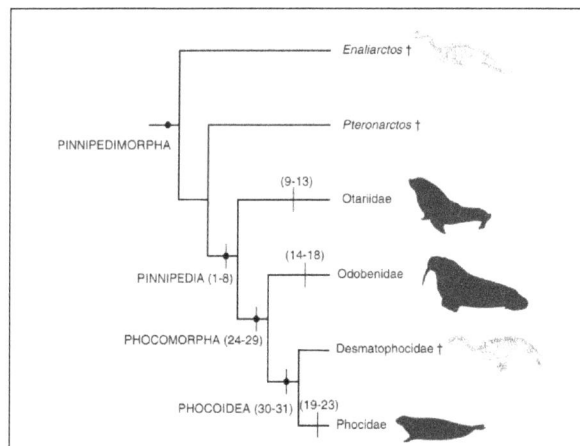

A cladogram depicting the relationships
of the major clades of pinnipeds.

- Greater and lesser humeral tubercles enlarged: Pinnipeds are distinguished from terrestrial carnivores by having strongly developed tubercles (rounded prominences) on the proximal end of the humerus (upper arm bone).

- Deltopectoral crest of humerus strongly developed: The crest on the shaft of the humerus for insertion of the deltopectoral muscles in pinnipeds is strongly developed in contrast to the weak development observed in terrestrial carnivores.

- Short and robust humerus: The short and robust humerus of pinnipeds is in contrast to the long, slender humerus of terrestrial carnivores.

- Digit I on the hand emphasized: In the hand of pinnipeds the first digit (thumb equivalent) is elongated, whereas in other carnivores the central digits are the most strongly developed.

- Digit I and V on the foot emphasized: Pinnipeds have elongated side toes (digits I and V, equivalent to the big toe and little toe) of the foot, whereas in other carnivores the central digits are the most strongly developed.

Pinniped Affinities

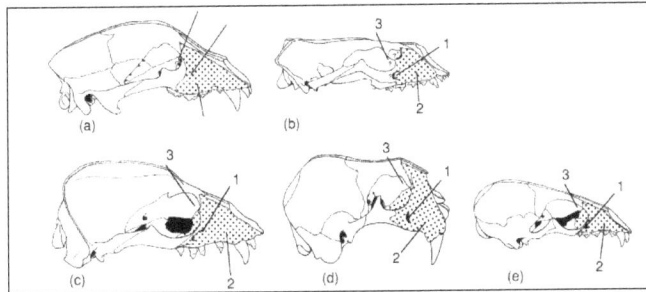

In the above figure, lateral views of the skulls of representative pinnipeds and a generalized terrestrial arctoid. (a) Bear,Ursus americanus. (b) Fossil pinnipedimorph, Enaliarctos mealsi. (c) Otariid,Zalophus californianus. (d) Walrus, Odobenus rosmarus. (e) Phocid, Monachus schauinslandi, illustrating pinniped synapomorphies. Character numbers: 1 = large infraorbital foramen; 2 = maxilla (stippled) makes a significant contribution to the orbital wall; 3 = lacrimal absent or fusing early and does not contact jugal.

Since the name Pinnipedia was first proposed by Illiger in 1811, there has been debate on the relationships of pinnipeds to one another and to other mammals. Two hypotheses have been proposed. The monophyletic hypothesis proposes that the three pinniped families share a single common evolutionary origin. The diphyletic view calls for the origin of pinnipeds from two carnivore lineages, the alliance of odobenids and otariids being somewhere near ursids (bears) and a separate origin for phocids from the mustelids (weasels, skunks, otters, and kin).

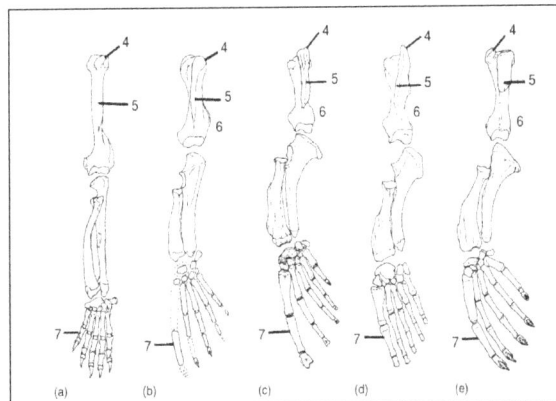

In the above figure, left forelimbs of representative pinnipeds (b–e) and a generalized terrestrial arctoid (a) in dorsal view illustrating pinniped synapomorphies. Labels as in figure plus character numbers: 4 = greater and lesser humeral tubercles enlarged; 5 = deltopectoral crest of humerus strongly developed; 6 = short, robust humerus; 7 = digit I on manus emphasized.

Traditionally, morphological and paleontological evidence supported pinniped diphyly. On the basis of his reevaluation of the morphological evidence, Wyss argued in favor of a return to the single origin interpretation. This hypothesis of pinniped monophyly has received considerable support from both morphological and biomolecular studies.

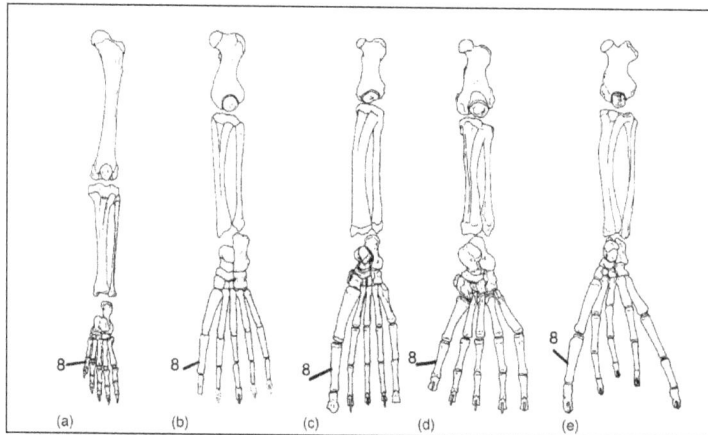

Left hind limbs of representative pinnipeds (b–e) and a generalized
terrestrial arctoid (a) in dorsal view illustrating pinniped synapomorphies.
Labels as in figure Character number: 8 = digit I and V on the foot emphasized.

All recent workers, on the basis of both molecular and morphologic data, agree that the closest relatives of pinnipeds are arctoid carnivores, which include procyonids (raccoons and their allies), mustelids, and ursids, although which specific arctoid group forms the closest alliance with pinnipeds is still disputed. There is evidence to support a mustelid, ursid and ursid-mustelid ancestry.

Although both morphological and molecular data support pinniped monophyly there is still disagreement on relationships among pinnipeds. Most of the controversy lies in the debate as to whether the walrus is most closely related to phocids or to otariids. Some recent morphologic evidence for extant pinnipeds unites the walrus and phocids as sister groups. An alternative view based mostly on molecular data but with support from total evidence analyses supports an alliance between the walrus and otariids.

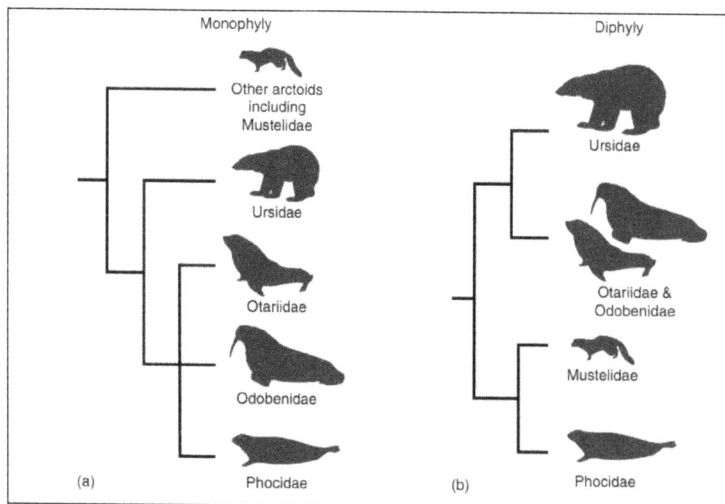

Alternative hypotheses for relationships among pinnipeds.
(a) Monophyly with ursids as the closest pinniped relatives. (b) Diphyly in which phocids
and mustelids are united as sister taxa, as are otariids, odobenids, and ursids.

Early "Pinnipeds"

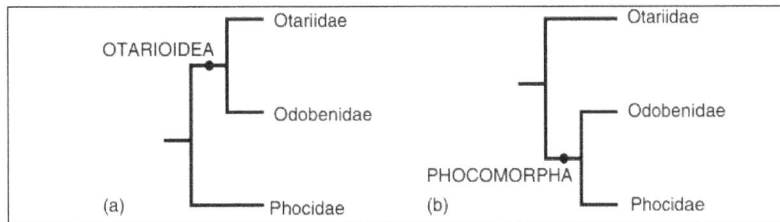

Alternative hypotheses for position of the walrus.
(a) "Otarioidea" clade. (b) Phocomorpha clade.

An understanding of the evolution of early "pinnipeds" necessitates a knowledge of certain fossil taxa. The earliest diverging lineage of "pinnipeds"actually are members of the Pinnipedimorpha clade and appear to have originated in the eastern North Pacific (Oregon) during the late Oligocene. The earliest known pinnipedimorph, Enaliarctos, is represented by five species. The ancestral pinnipedimorph dentition, exemplified by E. barnesi and E. mealsi, is heterodont, with large blade-like cusps on the upper cheekteeth well-adapted for shearing. These dental features together with those from the skull (when compared with terrestrial carnivores) indicate closest similarity in terms of derived characters with archaic bears.

Other species of the genus Enaliarctos show a trend toward the decreasing shearing function of the cheekteeth (e.g., reduction in the number and size of cusps). These dental trends herald the development of simple peg-like, or homodont, cheekteeth characteristic of most living pinnipeds. The latest record of Enaliarctos is along the Oregon coast from rocks of 25–18 Ma in age. An "enaliarctine" pinniped also has been reported from the western North Pacific (Japan) in rocks of late early Miocene, although the specimen needs further study before its taxonomic assignment can be confirmed.

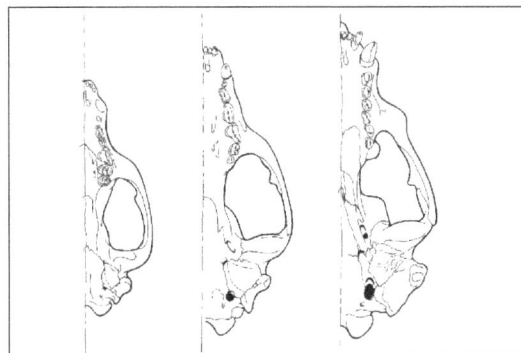

Skulls and dentitions of representative pinnipeds and a generalized terrestrial arctoid in ventral view.
(a) Archaic bear, Pachcynodon (Oligocene, France). (b) Fossil pinnipedimorph, Enaliarctos mealsi
(early Miocene). (c) Modern otariid, Arctocephalus (Recent, South Atlantic).

The pinnipedimorph E. mealsi is represented by a nearly complete skeleton collected from the Pyramid Hill Sandstone Member of the Jewett Sand in central California. The entire animal is estimated at 1.4–1.5 m in length and between 73 and 88 kg in weight, roughly the size and weight of a small male harbor seal.

Considerable lateral and vertical movement of the vertebral column was possible in E. mealsi. Also, both the fore- and hind limbs were modified as flippers and used in aquatic locomotion.

Several features of the hind limb suggest that E. mealsiwas highly capable of maneuvering on land and probably spent more time near the shore than extant pinnipeds.

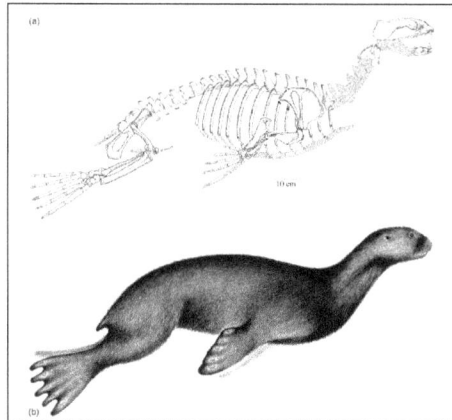

The pinnipedimorph,Enaliarctos mealsi.
(a)Skeletal reconstruction. (b)Life restoration. Total estimated length,
snout to tail, 1.4–1.5 m. Shaded areas are unpreserved bones.

A later diverging lineage of fossil pinnipeds more closely allied with pinnipeds than with Enaliarctosis Pteronarctos and Pacificotaria from the early-middle Miocene (19–15 Ma) of coastal Oregon. A striking osteological feature in all pinnipeds is the geometry of bones that comprise the orbital region. In Pteronarctos, the first evidence of the uniquely developed maxilla is seen. Also, in Pteronarctosthe lacrimal is greatly reduced or absent, as it is in pinnipeds. A shallow pit on the palate between the last premolar and the first molar, seen in Pteronarctos and pinnipeds, is indicative of a reduced shearing capability of the teeth and begins a trend toward homodonty.

Modern Pinnipeds

Family Otariidae: Sea Lions and Fur Seals

Of the two groups of seals, the otariids are characterized by the presence of external ear flaps, or pinnae, and for this reason they are sometimes called eared seals.

Representative otariids. (a) Southern sea lion, Otaria byronia and
(b) South African fur seal, Arctocephalus pusillus, illustrating pinna.
Note also the thick, dense fur characteristic of fur seals.

Another characteristic feature of otariids that can be used to distinguish them from phocids is their method of movement on land. Otariids can turn their hindflippers forward and use them to walk. Otariids generally are smaller than most phocids and are shallow divers targeting fast swimming fish as their major food source. The eared seals and sea lions, Family Otariidae, can be diagnosed as a monophyletic group by several osteological and soft anatomical characters as follows:

- Frontals extend anteriorly between nasals: In otariids, the suture between the frontal and nasal bones is W-shaped (i.e., the frontals extend between the nasals). In other pinnipeds and terrestrial carnivores, the contact between these bones is either transverse (terrestrial carnivores and walruses) or V-shaped (phocids).

- Supraorbital process of the frontal bone is large and shelf-like, especially among adult males. In otariids, the unique size and shape of the supraorbital process, located above the eye orbit, readily distinguishes them from other pinnipeds. The supraorbital process is absent in phocids and the modern walrus.

- Secondary spine subdivides the supraspinous fossa of the scapula: A ridge subdividing the supraspinous fossa of the scapula (shoulder blade) is present in otariids but not in walruses or phocids.

- Uniformly spaced pelage units: In otariids, pelage units (a primary hair and its surrounding secondaries) are spaced uniformly. In odobenids and phocids, the units are arranged in groups of two to four or in rows.

- Trachea has an anterior bifurcation of the bronchi: In odobenids and phocids, the trachea divides into two primary bronchi immediately outside the lung. In otariids, this division occurs more anteriorly, closer to the larynx and associated structures.

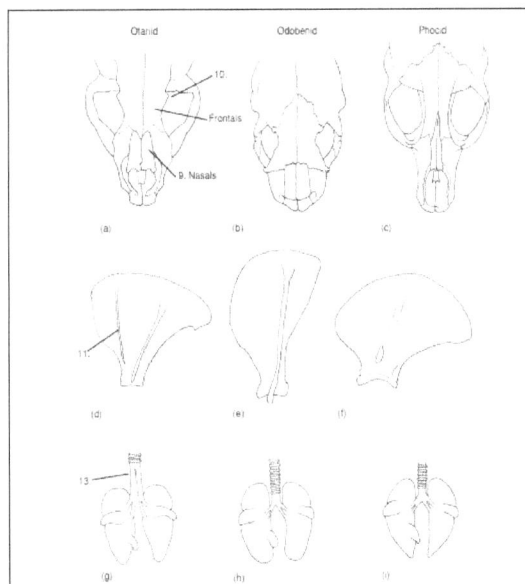

In the above figure, otariid synapomorphies. Character numbers: (a–c) Skulls in dorsal view: 9 = frontals extend anteriorly between nasals, contact between these bones is transverse (walrus) or V-shaped (phocids); 10 = supraorbital process of the frontal bone is large and shelf-like, this process is absent in the modern walrus and phocids. (d–f) Left scapulae in medial view: 11 = secondary spine subdividing the supraspinous fossa of the scapula, this ridge is absent in the walrus and phocids. (g–i) Lungs in ventral view: 13 = trachea has an anterior bifurcation of the bronchi. This division occurs immediately outside the lungs in the walrus and phocids.

The Otariidae often are divided into two subfamilies, the Otariinae (sea lions) and the Arcto-cephalinae (fur seals). Five genera and species of sea lions are recognized: Eumetopias, Neophoca, Otaria, Zalophus, and Phocarctos. Sea lions are characterized and readily distinguished from fur seals by their sparse pelage. The fur seals, named for their thick dense fur, are divided into two genera. Arctocephalus (generic name means bear head), or southern fur seals, live mostly in the southern hemisphere and a single species of northern fur seal, Callorhinus ursinus (generic name means beautiful nose), inhabits the northern hemisphere. Relationships among the otariids based on morphology indicate that only the Otariinae are monophyletic with a sister group relationship suggested with Arctocephalus. Callorhinus and the extinct taxon Thalassoleon are positioned as sequential sister taxa to this clade. Another recent analysis suggested that both subgroups were monophyletic.

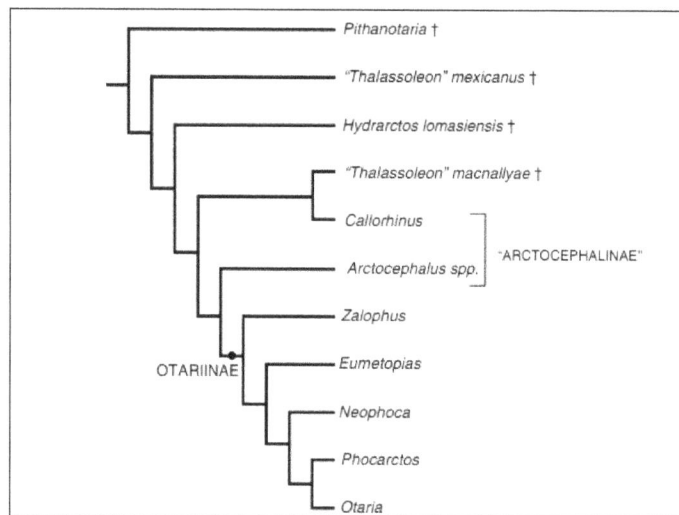

Phylogeny of the Otariidae based on morphologic data showing
monophyletic Otariinae and paraphyletic "Arctocephalinae" with † = extinct taxa.

Molecular sequence data revealed paraphyly among both fur seals and sea lions. New Zealand fur seal (Arctocephalus forsteri) and the northern fur seal (Callorhinus ursinus), both arctocephalines, are separated from each other by two sea lion lineages (Steller's sea lion, Eumetopias jubatus, and Hooker's sea lion, Phocarctos hookeri), and the two sea lions are no more closely related to each other than they are to other otariid taxa (i.e., the arctocephalines). A different arrangement among otariids is suggested by Árnason et al. but a limited number of species were sampled. Their study supports an alliance between Arctocephalus forsteri and the Antarctic fur seal, Arctocephalus gazella, and unites Steller's sea lion, Eumetopias, and the California sea lion, Zalophus. In addition to the extant fur seal genera Callorhinus and Arctocephalus, several extinct otariids are known. The earliest otariid is Pithanotaria starri from the late Miocene (11 Ma) of California. It is a small animal characterized by double rooted cheekteeth and a postcranial skeleton that allies it with other ota-riids. A second extinct late Miocene taxon (8–6 Ma), Thalassoleon, recently reviewed by Deméré and Berta (in press) is represented by three species: T. mexicanus from Cedros Island, Baja Cali-fornia, Mexico, and southern California; T. macnallyae from California; and T. inouei from central Japan. Thalassoleon is distinguished from Pithanotaria by its larger size and in lacking a thickened ridge of tooth enamel at the base of the third upper incisor. A single extinct species of northern fur seal, Callorhinus gilmorei, from the late Pliocene in southern California and Mexico and Japan has been described on the basis of a partial mandible, some teeth, and postcranial bones. Several

species of the southern fur seal genus Arctocephalus are known from the fossil record. The earliest known taxa are A. pusillus (South Africa) and A. townsendi (California) from the late Pleistocene.

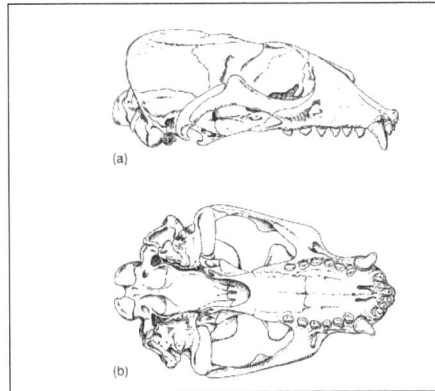

Skull of an early otariid, Thalassoleon mexicanus, from the late Miocene
of western North America in (a) lateral and (b) ventral views. Original 25 cm long.

The fossil record of sea lions is not well known. Only the late Pleistocene occurrences of Otaria byronia from Brazil and Neophoca palatina from New Zealand can be considered reliable.

Family Odobenidae: Walruses

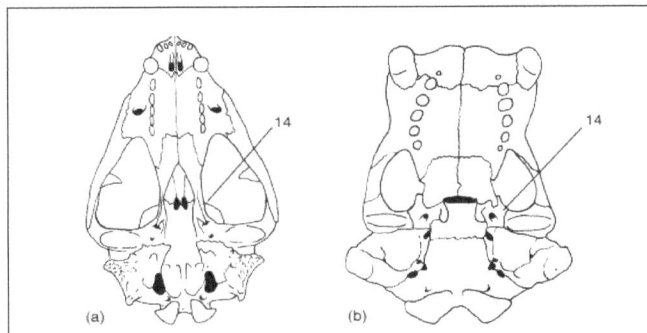

A walrus synapomorphy. Skull of an (a) otariid and
(b) walrus in ventral view illustrating differences in the pterygoid region. Character
number: 14 = broad, thick pterygoid strut; in the otariid the pterygoid strut is narrow.

Arguably the most characteristic feature of the modern walrus, Odobenus rosmarus, is a pair of elongated ever-growing upper canine teeth (tusks) found in adults of both sexes. A rapidly improving fossil record indicates that these unique structures evolved in a single lineage of walruses and that "tusks do not a walrus make." The modern walrus is a large-bodied shallow diver that feeds principally on benthic invertebrates, especially molluscs. Two subspecies of Odobenus rosmarus are usually recognized, Odobenus r. rosmarusfrom the North Atlantic and Odobenus r. divergens from the North Pacific. A population from the Laptev Sea has been described as a third subspecies, Odobenus. r. laptevi. Monophyly of the walrus family, the Odobenidae, is based on four unequivocal synapomorphies:

- Pterygoid strut broad and thick: The pterygoid strut is the horizontally positioned expanse of palatine, alisphenoid, and pterygoid lateral to the internal nares and hamular process. Basal pinnipedimorphs are characterized by having a narrow pterygoid strut, which in walruses is broad with a ventral exposure of the alisphenoid and pterygoid.

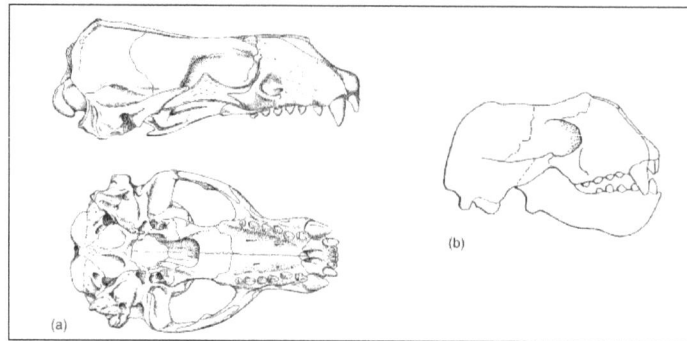

Skulls of fossil odobenids. (a) Lateral and ventral views of Imagotaria
downsi from the Miocene of western North America, Original 30 cm long.
(b) Lateral view of Protodobenus japonicus from the early Pliocene of Japan, Original 25 cm.

- P4 protocone shelf strong and posterolingually placed with convex posterior margin: In basal walruses (i.e., Proneotherium, Imagotaria, and Prototaria) the P4 protocone is a posteromedially placed shelf. That differs from Enaliarctos, which has a anterolingually placed protocone shelf. In later diverging walruses (i.e., Dusignathus and odobenines) the protocone shelf is greatly reduced or absent.

- M1 talonid heel absent: The condition in walruses (absence of talonind heel) differs from other pinnipedimorphs in which a distinct cusp, the hypoconulid, is developed on the talonid heel.

- Calcaneum with prominent medial tuberosity: In basal pinnipedimorphs, otariids and phocids, the calacaneal tuber is straight-sided. In walruses a prominent medial protuberance is developed on the proximal end of the calcaneal tuber.

Morphological study of evolutionary relationships among walruses has identified two monophyletic groups. The Dusignathinae includes the extinct genera Dusignathus, Gomphotaria, Pontolis, and Pseudodobenus. The Odobenidae includes in addition to the modern walrus, Odobenus, the extinct genera Aivukus, Alachtherium, Gingimanducans, Prorosmarus, Protodobenus, and Valenictus. Dusignathine walruses developed enlarged upper and lower canines, whereas odobenines evolved only the enlarged upper canines seen in the modern walrus.

At the base of walrus evolution are Proneotherium and Prototaria, from the middle Miocene (16–14 Ma) of the eastern North Pacific. Other basal odobenids include Neotherium and Imagotaria from the middle-late Miocene of the eastern North Pacific. These archaic walruses are characterized by unenlarged canines and narrow, multiple rooted premolars with a trend toward molarization, adaptations suggesting retention of the fish diet hypothesized for archaic pinnipeds rather than the evolution of the specialized mollusc diet of the modern walrus. The dusignathine walrus, Dusignathus santacruzensis, and the odobenine walrus, Aivukus cedroensis, first appeared in the late Miocene of California and Baja California, Mexico. Early diverging odobenine walruses are now known from both sides of the Pacific in the early Pliocene. Prorosmarus alleniis known from the eastern United States (Virginia) and Protodobenus japonicus from Japan. A new species of walrus, possibly the most completely known fossil odobenine walrus, Valenictus chulavistensis, was described by Deméré as being closely related to modern Odobenusbut distinguished from it in having no teeth in the lower jaw and lacking all upper postcanine teeth. The toothlessness (except for tusks) of Valenictus is unique among pinnipeds but parallels the condition seen in modern suction feeding whales and the narwhal. Remains of the modern walrus Odobenus date back to the early Pliocene of Belgium; this taxon appeared approximately 600,000 years ago in the Pacific.

Family Phocidae: Seals

The second major grouping of living seals, the phocids, often are referred to as the earless seals for their lack of visible ear pinnae, a characteristic that readily distinguishes them from otariids. Another characteristic phocid feature is their method of movement on land. The phocids are unable to turn their hindflippers forward and progression over land is accomplished by undulations of the body. Other characteristics of phocids include their larger body size in comparison to otariids, averaging as much as 2 tons in elephant seal males. Several phocids, most notably the elephant seal and the Weddell seal, are spectacular divers that feed on pelagic, vertically migrating squid and fish at depths of 1000 m or more.

Wyss (1988) reviewed the following characters that support monophyly of the Family Phocidae:

- Lack the ability to draw the hind limbs forward under the body due to a massively developed astragalar process and greatly reduced calcaneal tuber. The phocid astragalus (ankle bone) is distinguished by a strong posteriorly directed process over which the tendon of the flexor hallucis longus passes. The calcaneum (one of the heel bones) of phocids is correspondingly modified. The calcaneal tuber is shortened and projects only as far as the process of the astragalus. This arrangement prevents anterior flexion of the foot, resulting in seals' inability to bring their hind limbs forward during locomotion on land.

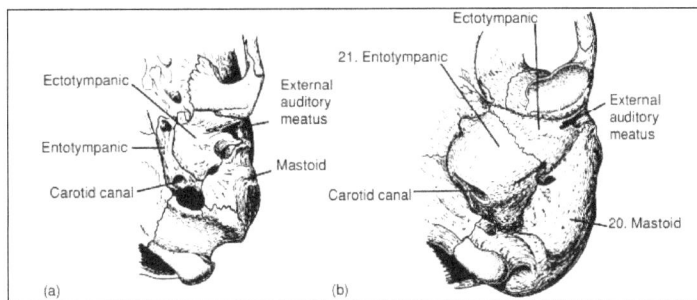

Phocid synapomorphies: Ventral view of ear region of (a) an otariid and (b) a phocid.
Character: 20 = pachyostotic mastoid bone—this is not the case in other pinnipeds; 21 = greatly inflated entotympanic bone—in other pinnipeds, this bone is flat or slightly inflated.

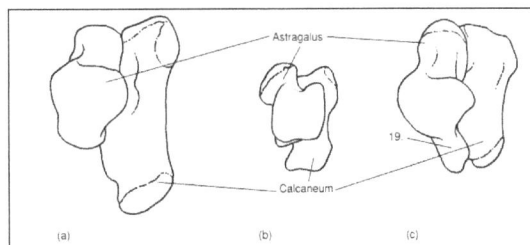

Phocid synapomorphies.

In the above figure, left astragalus (ankle) and calcaneum (heel) of (a) an otariid, (b) a walrus, and (c) a phocid. Character numbers: 19 = lack the ability to draw the hind limbs under the body due to a massively developed astragalar process and greatly reduced calcaneal tuber; these modifications do not occur in other pinnipeds.

- Pachyostic mastoid region: In phocids, the mastoid (ear) region is composed of thick, dense bone (pachyostosis), which is not the case in otariids or the walrus.

- Greatly inflated entotympanic bone: In phocids, the entotympanic bone (one of the bones forming the earbone or tympanic bulla) is inflated. In other pinnipeds, the entotympanic bone is either flat or slightly inflated.

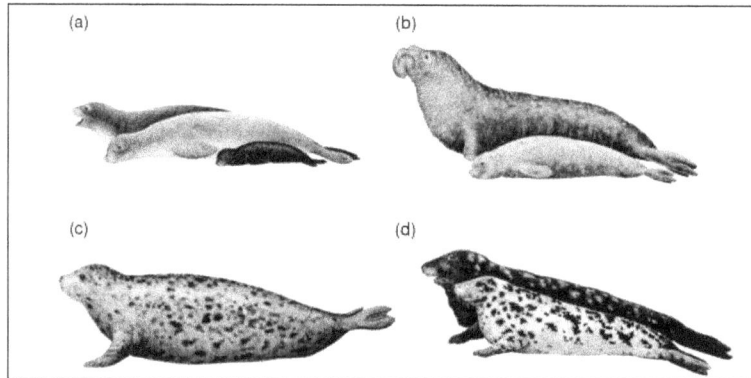

Representative "monachines"(a) Hawaiian monk seal,Monachus schauinslandi;(b) Northern elephant seal, Mirounga angustirostris, and phocines; (c) Harbor seal, Phoca vitulina; and (d) grey seal, Halichoerus grypus.

- Supraorbital processes completely absent: Phocids differ from other pinnipeds in the complete absence of the supraorbital process of the frontal.

- Strongly everted ilia: Living phocines, except Erignathus, are characterized by a lateral eversion (outward bending) of the ilium (one of the pelvic bones) accompanied by a deep lateral excavation. In other pinnipeds and terrestrial carnivores, the anterior termination of the ilium is simple and not everted or excavated.

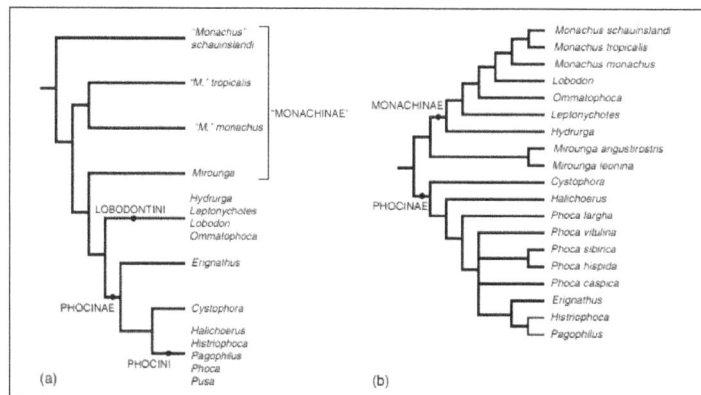

Alternative phylogenies for the Phocidae based on morphologic data. (a) From Wyss (1988) and Berta and Wyss (1994). (b) From Bininda-Emonds and Russell (1996).

Traditionally, phocids have been divided into two to four major subgroupings, monachines (monk seals), lobodontines (Antarctic seals), cystophorines (hooded and elephant seals), and phocines (remaining Northern Hemisphere seals). Based on morphologic data, Wyss argued for the monophyly of only one of these groups, the Phocinae, composed of Erignathus and Cystophora plus the tribe Phocini, consisting of Halichoerus, Histriophoca, Pagophilus, Phoca, and Pusa. According to Wyss, both the Monachinae"and the genus "Monachus" are paraphyletic with "Monachus," in turn representing the outgroup to the elephant seals,Mirounga, and the lobodontines (including the Weddell seal, Leptonychotes; crabeater seal, Lobodon; leopard seal, Hydrurga; and the Ross seal, Ommatophoca). Another morphology-based study found reasonable support for both the Monachinae and Phocinae, although with differing relationships among the taxa within each group. The basal position of Monachus and Erignathus, in the Monachinae and Phocinae, respectively, was not supported. Instead, both taxa were recognized as later diverging members of their respective subfamilies rendering the Lobodontini and Phocini as paraphyletic clades.

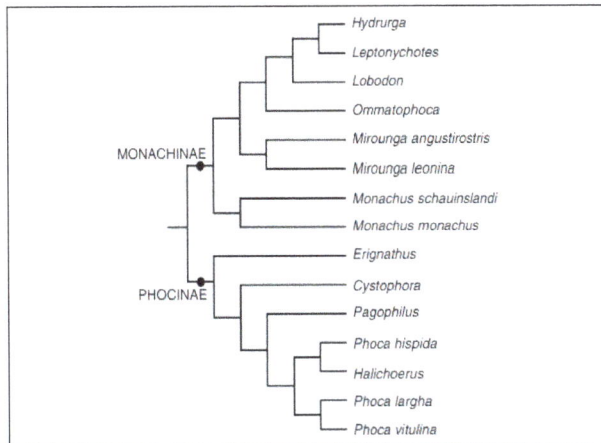

Phylogeny of the Phocidae based on molecular data.

EVOLUTION OF CETACEANS

The evolution of cetaceans is thought to have begun in the Indian subcontinent, from even-toed ungulates 50 million years ago, over a period of at least 15 million years. Cetaceans are fully aquatic marine mammals belonging to the order Artiodactyla, and branched off from other artiodactyls around 50 mya (million years ago). Cetaceans are thought to have evolved during the Eocene or earlier, sharing a closest common ancestor with hippopotamuses. Being mammals, they surface to breathe air; they have 5 finger bones (even-toed) in their fins; they nurse their young; and, despite their fully aquatic life style, they retained many skeletal features from their terrestrial ancestors. Discoveries starting in the late 1970s in Pakistan revealed several stages in the transition of cetaceans from land to sea.

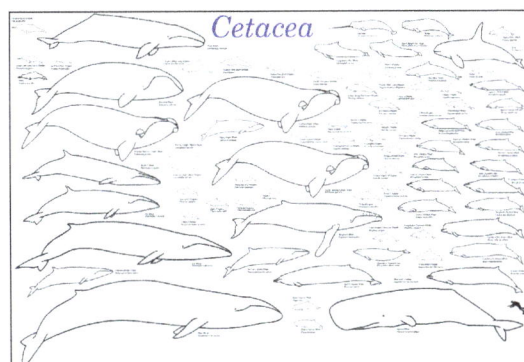

Species of the infraorder Cetacea.

The two modern parvorders of cetaceans – Mysticeti (baleen whales) and Odontoceti (toothed whales) – are thought to have separated from each other around 28-33 million years ago in a second cetacean radiation, the first occurring with the archaeocetes. The adaptation of animal echolocation in toothed whales distinguishes them from fully aquatic archaeocetes and early baleen whales. The presence of baleen in baleen whales occurred gradually, with earlier varieties having very little baleen, and their size is linked to baleen dependence (and subsequent increase in filter feeding).

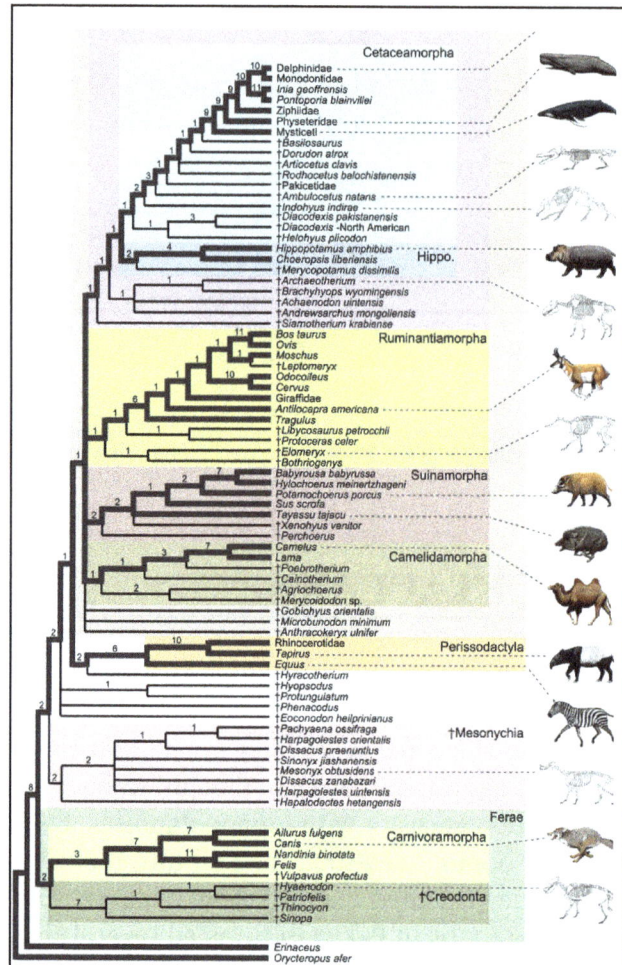

Cladogram showing the position of Cetacea within Artiodactylamorpha.

The aquatic lifestyle of cetaceans first began in the Indian subcontinent from even-toed ungulates 50 million years ago, over a period of at least 15 million years, however a jawbone discovered in Antarctica may reduce this to 5 million years. Archaeoceti is an extinct parvorder of Cetacea containing ancient whales. The traditional hypothesis of cetacean evolution, first proposed by Van Valen in 1966, was that whales were related to the mesonychids, an extinct order of carnivorous ungulates (hoofed animals) that resembled wolves with hooves and were a sister group of the artiodactyls (even-toed ungulates). This hypothesis was proposed due to similarities between the unusual triangular teeth of the mesonychids and those of early whales. However, molecular phylogeny data indicates that whales are very closely related to the artiodactyls, with hippopotamuses as their closest living relative. Because of this, cetaceans and hippopotamuses are placed in the same suborder, Whippomorpha.

Cetartiodactyla (formed from the words Cetacea and Artiodactyla) is a proposed name for an order containing both cetaceans and artiodactyls. However, the earliest anthracotheres, the ancestors of hippos, do not appear in the fossil record until the Middle Eocene, millions of years after Pakicetus, the first known whale ancestor, appeared during the Early Eocene, implying the two groups diverged well before the Eocene. Since molecular analysis identifies artiodactyls as being very closely related to cetaceans, mesonychids are probably an offshoot from Artiodactyla, and cetaceans did not derive directly from them, but that the two groups may share a common ancestor.

The molecular data are supported by the discovery of Pakicetus, the earliest archaeocete. The skeletons of Pakicetus show that whales did not derive directly from mesonychids. Instead, they are artiodactyls that began to take to the water soon after artiodactyls split from mesonychids. Archaeocetes retained aspects of their mesonychid ancestry (such as the triangular teeth) which modern artiodactyls, and modern whales, have lost. The earliest ancestors of all hoofed mammals were probably at least partly carnivorous or scavengers, and today's artiodactyls and perissodactyls became herbivores later in their evolution. Whales, however, retained their carnivorous diet because prey was more available and they needed higher caloric content in order to live as marine endotherms (warm-blooded). Mesonychids also became specialized carnivores, but this was likely a disadvantage because large prey was uncommon. This may be why they were out-competed by better-adapted animals like the hyaenodontids and later Carnivora.

Indohyus

Indohyus (Raoellidae cca 49–48 Ma).

Indohyus was a small chevrotain-like animal that lived about 48 million years ago in what is now Kashmir. It belongs to the artiodactyl family Raoellidae, and is believed to be the closest sister group of Cetacea. Indohyus is identified as an artiodactyl because it has two trochlea hinges, a trait unique to artiodactyls. Approximately the size of a raccoon or domestic cat, this omnivorous creature shared some traits of modern whales, most notably the involucrum, a bone growth pattern which is the diagnostic characteristic of any cetacean; this is not found in any other species. It also showed signs of adaptations to aquatic life, including dense limb bones that reduce buoyancy so that they could stay underwater, which are similar to the adaptations found in modern aquatic mammals such as the hippopotamus. This suggests a similar survival strategy to the African mousedeer or water chevrotain which, when threatened by a bird of prey, dives into water and hides beneath the surface for up to four minutes.

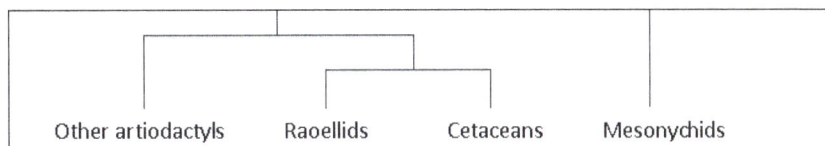

| Other artiodactyls | Raoellids | Cetaceans | Mesonychids |

Possible relationships between cetaceans and other ungulate groups.

Pakicetidae

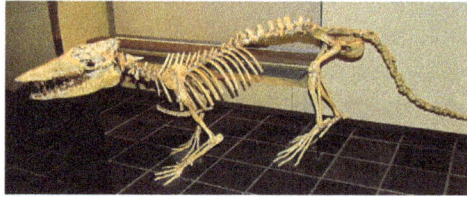

Pakicetus (Pakicetidae cca 49–48 Ma).

The pakicetids were digitigrade hoofed mammals that are thought to be the earliest known cetaceans, with Indohyus being the closest sister group. They lived in the early Eocene, around 50 million years ago. Their fossils were first discovered in North Pakistan in 1979, located at a river not far from the shores of the former Tethys Sea. After the initial discovery, more fossils were found, mainly in the early Eocene fluvial deposits in northern Pakistan and northwestern India. Based on this discovery, pakicetids most likely lived in an arid environment with ephemeral streams and moderately developed floodplains millions of years ago. By using stable oxygen isotopes analysis, they were shown to drink fresh water, implying that they lived around freshwater bodies. Their diet probably included land animals that approached water for drinking or some freshwater aquatic organisms that lived in the river. The elongated cervical vertebrae and the four, fused sacral vertebrae are consistent with artiodactyls, making Pakicetus one of the earliest fossils to be recovered from the period following the Cetacea/Artiodactyla divergence event.

Pakicetids are classified as cetaceans mainly due to the structure of the auditory bulla (ear bone), which is formed only from the ectotympanic bone. The shape of the ear region in pakicetids is highly unusual and the skull is cetacean-like, although a blowhole is still absent at this stage. The jawbone of pakicetids also lacks the enlarged space (mandibular foramen) that is filled with fat or oil, which is used in receiving underwater sound in modern cetaceans. They have dorsal orbits (eye sockets facing up), which are similar to crocodiles. This eye placement helps submerged predators observe potential prey above the water. According to a 2009 study, the teeth of pakicetids also resemble the teeth of fossil whales, being less like a dog's incisors, and having serrated triangular teeth, which is another link to more modern cetaceans. It was initially thought that the ears of pakicetids were adapted for underwater hearing, but, as would be expected from the anatomy of the rest of this creature, the ears of pakicetids are specialized for hearing on land. However, pakicetids were able to listen underwater by using enhanced bone conduction, rather than depending on the tympanic membrane like other land mammals. This method of hearing did not give directional hearing underwater.

Pakicetids have long thin legs, with relatively short hands and feet which suggest that they were poor swimmers. To compensate for that, their bones are unusually thick (osteosclerotic), which is probably an adaptation to make the animal heavier to counteract the buoyancy of the water. According to a 2001 morphological analysis by Thewissen et al., pakicetids display no aquatic skeletal adaptation; instead they display adaptations for running and jumping. Hence pakicetids were most likely aquatic waders.

Ambulocetidae

Ambulocetus, which lived about 49 million years ago, was discovered in Pakistan in 1994. They were vaguely crocodile-like mammals, possessing large brevirostrine jaws. In the Eocene, ambulocetids inhabited the bays and estuaries of the Tethys Sea in northern Pakistan. The fossils of ambulocetids are always found in near-shore shallow marine deposits associated with abundant marine plant fossils and littoral mollusks. Although they are found only in marine deposits, their oxygen isotope values indicate that they consumed a range of water with different degree of salinity, with some specimens having no evidence of sea water consumption and others that did not ingest fresh water at the time when their teeth were fossilized. It is clear that ambulocetids tolerated a wide range of salt concentrations. Hence, ambulocetids represent a transition phase of cetacean ancestors between fresh water and marine habitat.

Ambulocetus (Ambulocetidae cca 49–48 Ma).

The mandibular foramen in ambulocetids had increased in size, which indicates that a fat pad was likely to be housed in the lower jaw. In modern toothed whales, this fat pad in the mandibular foramen extends posteriorly to the middle ear. This allows sounds to be received in the lower jaw, and then transmitted through the fat pad to the middle ear. Similar to pakicetids, the orbits of ambulocetids are on the top of the skull, but they face more laterally than in pakicetids.

Ambulocetids had relatively long limbs with particularly strong hind legs, and they retained a tail with no sign of a fluke. The hindlimb structure of Ambulocetids shows that their ability to engage in terrestrial locomotion was significantly limited compared to that of contemporary terrestrial mammals, and likely did not come to land at all. The skeletal structures of the knee and ankle indicates that the motion of the hindlimbs was restricted into one plane. This suggests that, on land, propulsion of the hindlimbs was powered by the extension of dorsal muscles. They probably swam by pelvic paddling (a way of swimming which mainly utilizes their hind limbs to generate propulsion in water) and caudal undulation (a way of swimming which uses the undulations of the vertebral column to generate force for movements), as otters, seals and modern cetaceans do. This is an intermediate stage in the evolution of cetacean locomotion, as modern cetaceans swim by caudal oscillation (a way of swimming similar to caudal undulation, but is more energy efficient).

Recent studies showcase that ambulocetids were fully aquatic like modern cetaceans, possessing a similar thoracic morphology and being unable to support their weight on land. This suggests that complete abandonment of the land evolved much earlier among cetaceans than previously thought.

Remingtonocetidae

Kutchicetus (Remingtonocetidae cca 48 Ma).

Remingtonocetus (Remingtonocetidae cca 48 Ma).

Remingtonocetids lived in the Middle-Eocene in South Asia, about 49 to 43 million years ago. Compared to family Pakicetidae and Ambulocetidae, Remingtonocetidae was a diverse family found in north and central Pakistan and western India. Remingtonocetids were also found in shallow marine deposits, but they were obviously more aquatic than ambulocetidae. This is demonstrated by the recovery of their fossils from a variety of coastal marine environments, including near-shore and lagoonal deposits. According to stable oxygen isotopes analysis, most remingtonocetids did not ingest fresh water, and had hence lost their dependency on fresh water relatively soon after their origin.

The orbits of remingtonocetids faced laterally and were small. This suggests that vision was not an important sense for them. The nasal opening, which eventually becomes the blowhole in modern cetaceans, was located near the tip of the snout. The position of the nasal opening had remained unchanged since pakicetids. One of the notable features in remingtonocetids is that the semicircular canals, which are important for balancing in land mammals, had decreased in size. This reduction in size had closely accompanied the cetacean radiation into marine environments. According to a 2002 study done by Spoor et al., this modification of the semicircular canal system may represent a crucial 'point of no return' event in early cetacean evolution, which excluded a prolonged semi-aquatic phase.

Compared to ambulocetids, remingtonocetids had relatively short limbs. Based on their skeletal remains, remingtonocetids were probably amphibious cetaceans that were well adapted to swimming, and likely to swim by caudal undulation only.

Protocetidae

Rodhocetus (Protocetidae cca 45 Ma).

Protocetus (Protocetidae cca 45 Ma).

The protocetids form a diverse and heterogeneous group known from Asia, Europe, Africa, and North America. They lived in the Eocene, approximately 48 to 35 million years ago. The fossil remains of protocetids were uncovered from coastal and lagoonal facies in South Asia; unlike previous cetacean families, their fossils uncovered from Africa and North America also include open marine forms. They were probably amphibious, but more aquatic compared to remingtonocetids. Protocetids were the first cetaceans to leave the Indian subcontinent and disperse to all shallow subtropical oceans of the world. There were many genera among the family Protocetidae. Great variations in aquatic adaptations exist among them, with some probably able to support their weight on land, whereas others could not. Their supposed amphibious nature is supported by the discovery of a pregnant Maiacetus, in which the fossilised fetus was positioned for a head-first delivery, suggesting that Maiacetus gave birth on land. If they gave birth in the water, the fetus would be positioned for a tail-first delivery to avoid drowning during birth.

Unlike remingtonocetids and ambulocetids, protocetids have large orbits which are oriented laterally. Increasingly lateral-facing eyes might be used to observe underwater prey, and are similar to the eyes of modern cetaceans. Furthermore, the nasal openings were large and were halfway up the snout. The great variety of teeth suggests diverse feeding modes in protocetids. In both remingtonocetids and protocetids, the size of the mandibular foramen had increased. The large mandibular foramen indicates that the mandibular fat pad was present. However the air-filled sinuses that are present in modern cetaceans, which function to isolate the ear acoustically to enable better underwater hearing, were still not present. The external auditory meatus (ear canal), which is absent in modern cetaceans, was also present. Hence, the method of sound transmission that were present in them combines aspects of pakicetids and modern odontocetes (toothed whales). At this intermediate stage of hearing development, the transmission of airborne sound was poor due to the modifications of the ear for underwater hearing while directional underwater hearing was also poor compared to modern cetaceans.

Some protocetids had short, wide fore- and hindlimbs that were likely to have been used in swimming, but the limbs gave a slow and cumbersome locomotion on land. It is possible that some protocetids had flukes. However, it is clear that they were adapted even further to an aquatic lifestyle. In Rodhocetus, for example, the sacrum (a bone that, in land-mammals, is a fusion of five vertebrae that connects the pelvis with the rest of the vertebral column) was divided into loose vertebrae. However, the pelvis was still connected to one of the sacral vertebrae. The ungulate ancestry of these archaeocetes is still underlined by characteristics like the presence of hooves at the ends of the toes in Rodhocetus.

The foot structure of Rodhocetus shows that protocetids were predominantly aquatic. A 2001 study done by Gingerich et al. hypothesized that Rodhocetus locomoted in the oceanic environment similarly to how ambulocetids pelvic paddling, which was supplemented by caudal undulation. Terrestrial locomotion of Rodhocetus was very limited due to their hindlimb structure. It is thought that they moved in a way similar to how eared seals move on land, by rotating their hind flippers forward and underneath their body.

Basilosauridae

Archaeocetes (like this Basilosaurus) had a heterodont dentition.

Dorudon (Basilosauridae cca 35 Ma).

Squalodon (Squalodontidae, Odontoceti cca 25 Ma).

Basilosaurids and dorudontines lived together in the late Eocene around 41 to 35 million years ago, and are the oldest known obligate aquatic cetaceans. They were fully recognizable whales which lived entirely in the ocean. This is supported by their fossils usually found in deposits indicative of fully marine environments, lacking any freshwater influx. They were probably distributed throughout the tropical and subtropical seas of the world. Basilosaurids are commonly found in association with dorudontines, and were closely related to one another. The fossilised stomach contents in one basilosaurid indicates that it ate fish.

Kentriodon (Kentriodontidae.
Odontoceti cca 20 Ma).

Aulophyseter (Physeteridae,
Odontoceti cca 20 Ma).

Although they look very much like modern cetaceans, basilosaurids lacked the 'melon organ' that allows toothed whales to use echolocation. They had small brains; this suggests they were solitary and did not have the complex social structures of some modern cetaceans. The mandibular foramen of basilosaurids covered the entire depth of the lower jaw as in modern cetaceans. Their orbits faced laterally, and the nasal opening had moved even higher up the snout, closer to the position of the blowhole in modern cetaceans. Furthermore, their ear structures were functionally modern, with the insertion of air-filled sinuses between ear and skull. Unlike modern cetaceans, basilosaurids retained a large external auditory meatus.

Brygmophyseter (Physeteridae,
Odontoceti cca 15 Ma).

Aetiocetus (Aetiocetidae,
Mysticeti cca 27 Ma).

Both basilosaurids have skeletons that are immediately recognizable as cetaceans. A basilosaurid was as big as the larger modern whales, with genera like Basilosaurus reaching lengths of up to 60 ft (18 m) long; dorudontines were smaller, with genera like Dorudon reaching about 15 ft (4.6 m) long. The large size of basilosaurids is due to the extreme elongation of their lumbar vertebrae. They had a tail fluke, but their body proportions suggest that they swam by caudal undulation and that the fluke was not used for propulsion. In contrast, dorudontines had a shorter but powerful vertebral column. They too had a fluke and, unlike basilosaurids, they probably swam similarly to modern cetaceans, by using caudal oscillations. The forelimbs of basilosaurids were probably flipper-shaped, and the external hind limbs were tiny and were certainly not involved in locomotion. Their fingers, however, retained the mobile joints of their ambulocetid relatives. The two tiny but well-formed hind legs of basilosaurids were probably used as claspers when mating.

The pelvic bones associated with these hind limbs were not connected to the vertebral column as they were in protocetids. Essentially, any sacral vertebrae can no longer be clearly distinguished from the other vertebrae.

Janjucetus (Mammalodontidae, Mysticeti cca 25 Ma).

Cetotherium (Cetotheriidae, Mysticeti cca 18 Ma).

Both basilosaurids and dorudontines are relatively closely related to modern cetaceans, which belong to parvorders Odontoceti and Mysticeti. However, according to a 1994 study done by Fordyce and Barnes, the large size and elongated vertebral body of basilosaurids preclude them from being ancestral to extant forms. As for dorudontines, there are some species within the family that do not have elongated vertebral bodies, which might be the immediate ancestors of Odontoceti and Mysticeti. The other basilosaurids became extinct.

Evolution of Modern Cetaceans

Baleen Whales

Artistic impression of two Eobalaenoptera pursued by giant shark Carcharocles megalodon.

All modern baleen whales or mysticetes are filter-feeders which have baleen in place of teeth, though the exact means by which baleen is used differs among species (gulp-feeding within balaenopterids, skim-feeding within balaenids, and bottom plowing within eschrichtiids). The first members of both groups appeared during the middle Miocene. Filter feeding is very beneficial as it allows baleen whales to efficiently gain huge energy resources, which makes the large body size in modern varieties possible. The development of filter feeding may have been a result of worldwide environmental change and physical changes in the oceans. A large-scale change in ocean current and temperature could have contributed to the radiation of modern mysticetes. The earlier varieties of baleen whales, or "archaeomysticetes", such as Janjucetus and Mammalodon had very little baleen and relied mainly on their teeth.

There is also evidence of a genetic component of the evolution of toothless whales. Multiple mutations have been identified in genes related to the production of enamel in modern baleen whales. These are primarily insertion/deletion mutations that result in premature stop codons. It is hypothesized that these mutations occurred in cetaceans already possessing preliminary baleen structures, leading to the pseudogenization of a "genetic toolkit" for enamel production. Recent research has also indicated that the development of baleen and the loss of enamel-capped teeth both occurred once, and both occurred on the mysticete stem branch.

Generally it is speculated the four modern mysticete families have separate origins among the cetotheres. Modern baleen whales, Balaenopteridae (rorquals and humpback whale, Megaptera novaengliae), Balaenidae (right whales), Eschrichtiidae (gray whale, Eschrictius robustus), and Neobalaenidae (pygmy right whale, Caperea marginata) all have derived characteristics presently unknown in any cetothere and vice versa (such as a sagittal crest).

Toothed Whales

The adaptation of echolocation occurred when toothed whales (Odontoceti) split apart from baleen whales, and distinguishes modern toothed whales from fully aquatic archaeocetes. This happened around 34 million years ago in a second cetacean radiation. Modern toothed whales do not rely on their sense of sight, but rather on their sonar to hunt prey. Echolocation also allowed toothed whales to dive deeper in search of food, with light no longer necessary for navigation, which opened up new food sources. Toothed whales echolocate by creating a series of clicks emitted at various frequencies. Sound pulses are emitted, reflected off objects, and retrieved through the lower jaw. Skulls of Squalodon show evidence for the first hypothesized appearance of echolocation. Squalodon lived from the early to middle Oligocene to the middle Miocene, around 33–14 million years ago. Squalodon featured several commonalities with modern toothed whales: the cranium was well compressed (to make room for the melon, a part of the nose), the rostrum telescoped outward into a beak, a characteristic of the modern toothed whales that gave Squalodon an appearance similar to them. However, it is thought unlikely that squalodontids are direct ancestors of modern toothed whales.

Acrophyseter skull.

The first oceanic dolphins such as kentriodonts, evolved in the late Oligocene and diversified greatly during the mid-Miocene. The first fossil cetaceans near shallow seas (where porpoises inhabit) were found around the North Pacific; species like Semirostrum were found along California (in what were then estuaries). These animals spread to the European coasts and Southern Hemisphere only much later, during the Pliocene. The earliest known ancestor of arctic whales is

Denebola brachycephala from the late Miocene around 9–10 million years ago. A single fossil from Baja California indicates the family once inhabited warmer waters.

Ancient sperm whales differ from modern sperm whales in tooth count and the shape of the face and jaws. For example, Scaldicetus had a tapered rostrum. Genera from the Oligocene and Miocene had teeth in their upper jaws. These anatomical differences suggest that these ancient species may not have necessarily been deep-sea squid hunters like the modern sperm whale, but that some genera mainly ate fish. Contrary to modern sperm whales, most ancient sperm whales were built to hunt whales. Livyatan had a short and wide rostrum measuring 10 feet (3.0 m) across, which gave the whale the ability to inflict major damage on large struggling prey, such as other early whales. Species like these are collectively known as killer sperm whales.

Beaked whales consist of over 20 genera. Earlier variety were probably preyed upon by killer sperm whales and large sharks such as Megalodon. In 2008, a large number of fossil ziphiids were discovered off the coast of South Africa, confirming the remaining ziphiid species might just be a remnant of a higher diversity that has since gone extinct. After studying numerous fossil skulls, researchers discovered the absence of functional maxillary teeth in all South African ziphiids, which is evidence that suction feeding had already developed in several beaked whale lineages during the Miocene. Extinct ziphiids also had robust skulls, suggesting that tusks were used for male-male interactions.

Skeletal Evolution

Dolphins (aquatic mammals) and ichthyosaurs (extinct marine reptiles) share a number of unique adaptations for fully aquatic lifestyle and are frequently used as extreme examples of convergent evolution.

Modern cetaceans have internal, rudimentary hind limbs, such as reduced femurs, fibulas, and tibias, and a pelvic girdle. Indohyus has a thickened ectotympanic internal lip of the ear bone. This feature compares directly to that of modern cetaceans. Another similar feature was the composition of the teeth, which contained mostly calcium phosphate which is needed for eating and drinking by aquatic animals, though, unlike modern day toothed whales, they had a heterodont (more than one tooth morphology) dentition as opposed to a homodont (one tooth morphology present) dentition. Although they somewhat resembled a wolf, the fossils of pakicetids showed the eye sockets were much closer to the top of their head than that of other terrestrial mammals, but similar to the structure of the eyes in cetaceans. Their transition from land to water led to reshaping of the skull and food processing equipment because the eating habits were changing. Ultimately, the change in position of the eyes and limb bones is what led the pakicetids to become waders. The ambulocetids also began to develop long snouts, which is seen in current cetaceans. Their limbs (and hypothesized movement) were very similar to otters.

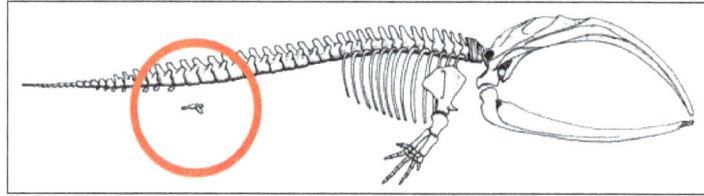

The skeleton of a Bowhead whale with the hind limb and pelvic bone structure circled in red. This bone structure stays internal during the entire life of the species.

Limblessness in cetaceans does not represent a regression of fully formed limbs nor the absence of limb bud initiation, but rather arrest of limb bud development. Limb buds develop normally in cetacean embryos.Limb buds progress to the condensation phase of early skeletogenesis, where nerves grow into the limb bud and the apical ectodermal ridge (AER), a structure that ensures proper limb development, appears functional. Occasionally, the genes that code for longer extremities cause a modern whale to develop miniature legs (atavism).

Pakicetus attocki skeleton.

Pakicetus had a pelvic bone most similar to that of terrestrial mammals. In later species, such as Basilosaurus, the pelvic bone, no longer attached to the vertebrae and the ilium, was reduced. Certain genes are believed to be responsible for the changes that occurred to the cetacean pelvic structure, such as BMP7, PBX1, PBX2, PRRX1, and PRRX2. The pelvic girdle in modern cetaceans were once thought to be vestigial structures that served no purpose at all. The pelvic girdle in male cetaceans is different in size compared to females, and the size is thought to be a result of sexual dimorphism. The pelvic bones of modern male cetaceans are more massive, longer, and larger than those of females. Due to the sexual dimorphism displayed, they were most likely involved in supporting male genitalia that remain hidden behind abdominal walls until sexual reproduction occurs.

Early archaeocetes such as Pakicetus had the nasal openings at the end of the snout, but in later species such as Rodhocetus, the openings had begun to drift toward the top of the skull. This is known as nasal drift. The nostrils of modern cetaceans have become modified into blowholes that allow them to break to the surface, inhale, and submerge with convenience. The ears began to move inward as well, and, in the case of Basilosaurus, the middle ears began to receive vibrations from the lower jaw. Today's modern toothed whales use their melon organ, a pad of fat, for echolocation.

Ongoing Evolution

Culture

Researchers pushed a pole with a conical sponge attached along the substrate to simulate sponging behavior by dolphins. They videotaped this experiment to learn what prey was available on the seafloor and why a sponge would be beneficial to foraging rather than echolocation.

Culture is group-specific behavior transferred by social learning. Tool use to aid with foraging is one example. Whether or not a dolphin uses a tool affects their eating behavior, which causes differences in diet. Also, using a tool allows a new niche and new prey to open up for that particular dolphin. Due to these differences, fitness levels change within the dolphins of a population, which further causes evolution to occur in the long run. Culture and social networks have played a large role in the evolution of modern cetaceans, as concluded in studies showing dolphins preferring mates with the same socially learned behaviors, and humpback whales using songs between breeding areas. For dolphins particularly, the largest non-genetic effects on their evolution are due to culture and social structure.

The population of Indo-Pacific bottlenose dolphins (Tursiops sp.) around Shark Bay of Western Australia can be divided into spongers and nonspongers. Spongers put sea sponges on their snout as a protective means against abrasions from sharp objects, stingray barbs, or toxic organisms. The sponges also help the dolphins target fish without swim bladders, since echolocation cannot detect these fish easily against a complex background. Spongers also specifically forage in deep channels, but nonspongers are found foraging in both deep and shallow channels. This foraging behavior is mainly passed on from mother to child. Therefore, since this is a group behavior being passed down by social learning, this tool use is considered a cultural trait.

In Shark Bay found the fatty acid analyses between the West and East Gulf populations to differ, which is due to the two areas having different food sources. However, when comparing data from within the West Gulf, the spongers vs. the nonspongers in the deep channels had very different fatty acid results even though they are in the same habitat. Nonspongers from deep and shallow channels had similar data. This suggests that sponging was the cause of the different data and not the deep vs. shallow channels. Sponging opened up a new niche for the dolphins and allowed them access to new prey, which caused long-term diet changes. By producing different food sources within a population, there is less intrapopulation competition for resources, showing character displacement. As a result, the carrying capacity increases since the entire population does not depend on one food source. The fitness levels within the population also change, thus allowing this culture to evolve.

Social Structure

Social structure forms groups with individuals that interact with one another, and this allows for cultural traits to emerge, exchange, and evolve. This relationship is especially seen in the bottlenose dolphin populations in southwestern Australia, which have been known to beg for food from fishermen. This begging behavior was spread through the population due to individual (dolphins spending time around boats) and social (dolphins spending time with other dolphins who express begging behavior) learning.

Culture can, however, impact social structure by causing behavior matching and assertive mating. Individuals within a certain culture are more likely to mate with individuals using the same behaviors rather than a random individual, thus influencing social groups and structure. For example, the spongers of Shark Bay preferentially stick with other spongers. Also, some bottlenose dolphins in Moreton Bay, Australia followed prawn trawlers to feed on their debris, while other dolphins in the same population did not. The dolphins preferentially associated with individuals with same behavior even though they all lived in the same habitat. Later on, prawn trawlers were no longer present, and the dolphins integrated into one social network after a couple of years.

Social networks can still affect and cause evolution on their own by impending fitness differences on individuals. According to a 2012 study, male calves had a lower survival rate if they had stronger bonds with juvenile males. However, when other age and sex classes were tested, their survival rate did not significantly change. This suggests that juvenile males impose a social stress on their younger counterparts. In fact, it has been documented that juvenile males commonly perform acts of aggression, dominance, and intimidation against the male calves. According to a 2010 study, certain populations of Shark Bay dolphins had varying levels of fitness and calf success. This is either due to social learning (whether or not the mother passed on her knowledge of reproductive ability to the calves), or due to the strong association between mother dolphins in the population; by sticking in a group, an individual mother does not need to be as vigilant all the time for predators.

Genetic studies conducted on Clymene dolphins (Stenella clymene) focused on their natural histories, and the results show that the origin of the species was actually an outcome of hybrid speciation. Hybridization between spinner dolphins (Stenella longirostris) and striped dolphins (Stenella coeruleoalba) in the North Atlantic was caused by constant habitat sharing of the two species. Relationships between these three species had been speculated according to notable resemblances between anatomies of the Clymene and the spinner dolphins, resulting in the former being regarded as subspecies of the latter until 1981, and the possibility of the Clymene dolphin as a hybrid between the spinner and the striped dolphins have come to question based on anatomical and behavioral similarities between these two species.

Environmental Factors

Genome sequences done in 2013 revealed that the Yangtze River dolphin, or "baiji" (Lipotes vexillifer), lacks single nucleotide polymorphisms in their genome. After reconstructing the history of the baiji genome for this dolphin species, researchers found that the major decrease in genetic diversity occurred most likely due to a bottleneck event during the last deglaciation event. During this time period, sea levels were rising while global temperatures were decreasing. Other historical climate events can be correlated and matched with the genome history of the Yangtze River dolphin as well. This shows how global and local climate change can drastically affect a genome, leading to changes in fitness, survival, and evolution of a species.

The European population of common dolphins (Delphinus delphis) in the Mediterranean have differentiated into two types: eastern and western. According to a 2012 study, this seems to be due to a recent bottleneck as well, which drastically decreased the size of the eastern Mediterranean population. Also, the lack of population structure between the western and eastern regions seems contradictory of the distinct population structures between other regions of dolphins. Even though the dolphins in the Mediterranean area had no physical barrier between their regions, they still differentiated into two types due to ecology and biology. Therefore, the differences between the eastern and western dolphins most likely stems from highly specialized niche choice rather than just physical barriers. Through this, environment plays a large role in the differentiation and evolution of this dolphin species.

The divergence and speciation within bottlenose dolphins has been largely due to climate and environmental changes over history. According to research, the divisions within the genus correlate with periods of rapid climate change. For example, the changing temperatures could cause the coast landscape to change, niches to empty up, and opportunities for separation to appear. In the Northeast Atlantic, specifically, genetic evidence suggests that the bottlenose dolphins have differentiated into coastal and pelagic types. Divergence seems most likely due to a founding event where a large group separated. Following this event, the separate groups adapted accordingly and formed their own niche specializations and social structures. These differences caused the two groups to diverge and to remain separated.

Two endemic, distinctive types of short-finned pilot whale, Tappanaga (or Shiogondou) the larger, northern type and Magondou the smaller, southern type, can be found along the Japanese archipelago where distributions of these two types mostly do not overlap by the oceanic front border around the easternmost point of Honshu. It is thought that the local extinction of long-finned pilot whales in the North Pacific in the 12th century could have triggered the appearance of Tappanaga, causing short-finned pilot whales to colonize the colder ranges of the long-finned variant. Whales with similar characteristics to the Tappanaga can be found along Vancouver Island and northern USA coasts as well.

EVOLUTION OF SIRENIAN

The closest living relatives of sirenians are Proboscidea (elephants). The Sirenia, the Proboscidea, the extinct Desmostylia, and probably the extinct Embrithopoda together make up a larger group called Tethytheria, whose members (as the name indicates) appear to have evolved from primitive hoofed mammals (condylarths) in the Old World along the shores of the ancient Tethys Sea. Together with Hyracoidea (hyraces), tethytheres seem to form a more inclusive group long referred to as Paenungulata. The Paenungulata and (especially) Tethytheria are among the least controversial groupings of mammalian orders and are strongly supported by most morphological and molecular studies. Their ancestry is remote from that of cetaceans or pinnipeds; sirenians reevolved an aquatic lifestyle independently of (though simultaneously with) cetaceans, ultimately displaying strong convergence with them in body form.

Early History, Anatomy and Mode of Life

Sirenians first appear in the fossil record in the early Eocene, and the order was already diverse by the middle Eocene. As inhabitants of rivers, estuaries, and nearshore marine waters, they were able to spread quickly along the coasts of the world's shallow tropical seas; in fact, the most primitive sirenian known to date (Prorastomus) was found not in the Old World but in Jamaica.

The earliest sea cows (families Prorastomidae and Proto-sirenidae, both confined to the Eocene) were pig-sized, four-legged amphibious creatures. By the end of the Eocene, with the appearance of the Dugongidae, sirenians had taken on their modern, completely aquatic, streamlined body form, featuring flipper-like front legs, no hind legs, and a powerful tail with a horizontal caudal fin, whose up-and-down movements propel them through the water, as in whales and dolphins. The last-appearing of the four sirenian families (Trichechi-dae) apparently arose from early dugongids in the late Eocene or early Oligocene. The sirenian fossil record now documents all the major stages of hindlimb and pelvic reduction from completely "terrestrial" morphology to the extremely reduced condition of the pelvis seen in modern manatees, thereby providing one of the most dramatic examples of evolutionary change to be seen among fossil vertebrates.

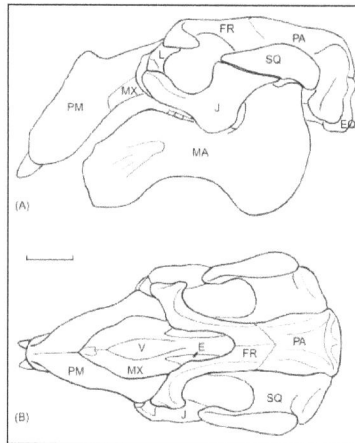

Simplified phytogeny of Sirenia, including only better-known genera.
The time scale (at left) is in millions of years. "Ghost lineages" (ancestral groups
undocumented by fossils) that span multiple epochal boundaries are shown as dashed lines.

From the outset, sirenians were herbivores and probably depended on seagrasses and other aquatic angiosperms (flowering plants) for food. To this day, almost all members of the order have remained tropical, marine, and eaters of angiosperms. No longer capable of locomotion on land, sirenians are born in the water and spend their entire lives there. Because they are shallow divers with large lungs, they have heavy skeletons, like a diver's weight belt, to help them stay submerged: their bones are both swollen (pachyostotic) and dense (osteosclerotic), especially the ribs, which are often found as fossils.

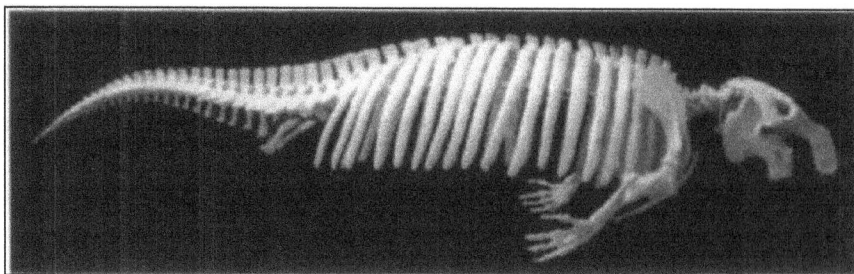

The sirenian skull is characterized by an enlarged and more or less downturned premaxillary rostrum, retracted nasal opening, absence of paranasal air sinuses, laterally salient zygomatic arches, and thick, dense parietals fused into a unit with the supraoccipital. Nasals and lacrimals tend to become reduced or lost, and in most forms the pterygoid processes are large and stout. The periotic is snugly enclosed by a socket in the squamosal and is fused with a ring-shaped tympanic. The mandibular symphysis is long, deep, laterally compressed, and typically fused and downturned; in all but prorastomids the mandibular foramen is enlarged to expose the dental capsule. Incisors, where present, are arranged in parallel, longitudinally aligned rows. In all but the most primitive taxa, the infraorbital and mental foramina are enlarged to accommodate the nerve and blood supply to the large, prehensile, vibrissae-studded lips, which are moved by muscular hydrostats.

Eocene sirenians, like Mesozoic mammals but in contrast to other Cenozoic ones, have five instead of four premolars, giving them a 3.1.5.3 dental formula. Whether this condition is truly a primitive retention in the Sirenia is still being debated. The fourth lower deciduous premolar (dp4) is trilobed, like that of many other ungulates; this raises the further unresolved question of whether the three following teeth (dp5, ml, and m2) are actually the homologues of the so-called ml-3 in other mammals.

Although the cheek teeth are relied on for identifying species in many other mammalian groups, they do not vary much in morphology among Sirenia but are almost always low-crowned (brachyodont) with two rows of large, rounded cusps (buno-bilophodont). (The most taxonomically informative parts of the sirenian skeleton are the skull and mandible, especially the frontal and other bones of the skull roof) Except for a pair of tusk-like first upper incisors seen in most species, front teeth (incisors and canines) are lacking in all but the earliest fossil sirenians, and cheek teeth in adults are commonly reduced in number to four or five on each side of each jaw: one or two deciduous premolars, which are never replaced, plus three molars. As described later, however, all three of the Recent genera have departed in different ways from this "typical" pattern.

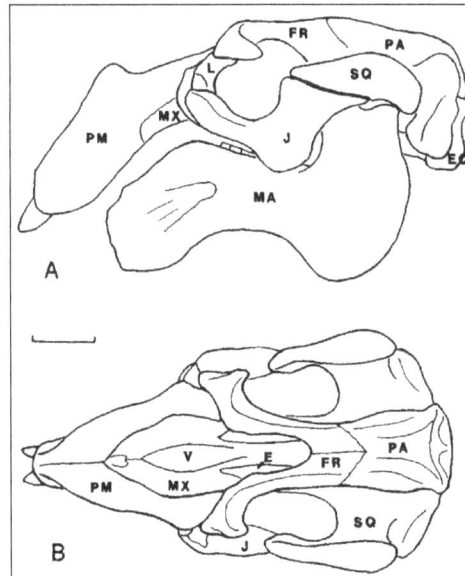

Skull of Crenatosiren olseni, an Oligocene dugongine dugongid, in (A) lateral and (B) dorsal views. Note the large incisor tusks in the premaxillae. E, ethmoid; EO, ex-occipital; FR, frontal;], jugal; L, lacimal; MA, mandible; MX, maxilla; PA, parietal; PM, premaxilla; SQ, squamosal; V, vomer. Scale bar: 5 cm.

Dugongidae

Dugongids comprise the vast majority of the species and specimens that make up the known fossil record of sirenians. The basal members ol tins very successful family are placed in the long-lived (Eocene-Pliocene) and cosmopolitan subfamily Halitheriinae. This paraphyletic group included the well-known fossil genera Halitherium and Metaxytherium, which were relatively unspecialized seagrass eaters.

Metaxytherium gave rise in the Miocene to the Hy-drodamalinae, an endemic North Pacific lineage that ended with Steller's sea cow (Hydrodamalis)—the largest sirenian that ever lived (up to 9 m or more in length) and the only one to adapt successfully to temperate and cold waters and a diet of marine algae. It was completely toothless, and its truncated, claw-like flippers, used for gathering plants and fending off from rocks, contained no finger bones (phalanges). It was hunted to extinction for its meat, fat, and hide circa a.d. 1768.

Another offshoot of the Halitheriinae, the subfamily Dugonginae, appeared in the Oligocene. Most dugongines were apparently specialists at digging out and eating the tough, buried rhizomes of seagrasses; for this purpose many of them had large, self-sharpening blade-like tusks. The modem Dugong is the sole survivor of this group, but it has reduced its dentition (the cheek teeth have only thin enamel crowns, which quickly wear off, leaving simple pegs of dentine) and has (perhaps for that reason) shifted its diet to more delicate seagrasses and ceased to use its tusks for digging.

Trichechidae

Trichechidae have a much less complete fossil record dian dugongids. Their definition has been broadened by Domning (1994) to include Miosireninae, a peculiar and little-known pair of genera that inhabited northwestern Europe in the late Oligocene and Miocene. Miosirenines had massively reinforced palates and dentitions that may have been used to crush shellfish. Such a diet in sirenians living around the North Sea seems less surprising when we consider that modern dugongs and manatees near the climatic extremes of their ranges are known to consume invertebrates in addition to plants.

Manatees in the strict, traditional sense are now placed in the subfamily Trichechinae. They first appeared in the Miocene, represented by Potamosiren from freshwater deposits in Colombia. Indeed, much of trichechine history was probably spent in South America, whence they spread to North America and Africa only in the Pliocene or Pleistocene.

During the late Miocene, manatees living in the Amazon basin evidently adapted to a diet of abrasive freshwater grasses by means of an innovation still used by their modem descendants; they continue to add on extra teeth to the molar series as long as they live, and as worn teeth fall out at the front, the whole tooth row slowly shifts forward to make room for new ones erupting at the rear. This type of horizontal tooth replacement has often been likened, incorrectly, to that of elephants, but the latter are limited to only three molars.

References

- Origins-marine-mammals: oceanadventures.co.za, Retrieved 6 July, 2019

- Choi, Charles (2014). "DNA Discovery Reveals Surprising Dolphin Origins". Retrieved 2 January 2016

- Sirenian-evolution-marine-mammals, marine-mammals: what-when-how.com, Retrieved 7 August, 2019

- M. G. Fitzgerald, Erich (2012). "Archaeocete-like jaws in a baleen whale". Biology Letters. 8 (1): 94–96. Doi:10.1098/rsbl.2011.0690. PMC 3259978. PMID 21849306

3

Cetaceans

Cetaceans are warm blooded aquatic mammals who have well-developed senses and a layer of fat under skin to maintain body heat in cold water. Some of the common cetaceans are whales, porpoises and dolphins. This chapter has been carefully written to provide an easy understanding of these cetaceans as well as their ecology and physiology.

Cetacean is a member of an entirely aquatic group of mammals commonly known as whales, dolphins, and porpoises. The ancient Greeks recognized that cetaceans breathe air, give birth to live young, produce milk, and have hair—all features of mammals. Because of their body form, however, cetaceans were commonly grouped with the fishes. Cetaceans are entirely carnivorous, although members of the order Sirenia (manatees, dugongs, and Steller's sea cow) were once referred to as the "herbivorous Cetacea." In the past cetaceans were important resources, but by the end of the 20th century their economic importance was almost solely due to whale watching, a tourist activity and major source of income for certain coastal regions of many countries.

Form and Function

Body Surface

Gray whale (Eschrichtius robustus).

The hair covering that is common to mammals is drastically reduced in cetaceans, likely because hair is a poor insulator when wet and increases drag during swimming. Hairs on cetaceans are restricted to the head, with isolated follicles occurring on the lower jaw and the snout. These are thought to be remnants of sensory whiskers (vibrissae). External pigmentation is important to many animals as a basis for individual recognition and species recognition. Hair defines the colour pattern of most mammals, but, because cetaceans have very little hair, the outer layer of skin (epidermis) produces their markings, most commonly in shades of black and white. The appearance of some cetaceans is affected by various organisms living on or in the skin. Examples include yellow algae that colour the lower body surface of blue whales (Balaenoptera musculus) and the variety of whitish organisms living on bodies of gray whales (Eschrichtius robustus) and right whales (family Balaenidae).

Locomotor Adaptations

The most noticeable adaptation of cetaceans to life in the water is their locomotive system. Because cetaceans descended from mammals that moved their limbs in a vertical plane rather than in a horizontal plane, they use vertical strokes when they swim, instead of horizontal strokes like a crocodile or fish. Cetaceans evolved from four-legged (quadruped) terrestrial animals, for which limbs played a primary role in movements, into virtually limbless aquatic creatures living in an environment where the back muscles are more important. Forelimbs are still present but are reduced to finlike flippers having shortened arm bones and no individual fingers. The hind limbs are lost entirely; only vestigial elements sometimes remain internally. Pelvic remnants occur in all cetacea but the dwarf and pygmy sperm whales. Flippers help to steer, while the back muscles, which are very large, drive the tail to propel the animal. Cetaceans have developed horizontal flukes that increase the propulsion area driven by the back muscles. Like fish, almost all cetaceans possess a dorsal fin that serves as a keel. The dorsal fin and flukes are composed of connective tissue, not bone. Other connective tissue, such as external ears, has been lost, and the male genitalia have moved internally.

Humpback whale (Megaptera novaeangliae) breaching.

Respiration

Normally, cetaceans breathe while moving through the water and spend only a short time at the surface, where they exhale in an explosive ventilation called a blow. The blow is expelled forcibly and can be compared to a cough. Cetaceans use up to 80 percent of their lung volume in a single breath, in contrast to humans, who use only 20 percent. The blow is visible because of water condensation and mucous particles; blows of blue whales are frequently more than 6 metres (20 feet) high. When a terrestrial mammal loses consciousness, it breathes reflexively, but breathing is not a reflex in cetaceans. Thus, when a cetacean loses consciousness, it does not breathe and quickly dies. For this reason, veterinarians had to perfect respirators before dolphins could be successfully anesthetized.

Circulation and Thermoregulation

Cetaceans, like all mammals, have a four-chambered heart with paired ventricles and auricles. The pattern of circulation is similar to that of other mammals, with the exception of a series of well-developed reservoirs for oxygenated blood called the rete mirabile, for "marvelous network."

These provide bypasses that enable cetaceans to isolate skeletal muscle circulation during diving while using the oxygen stored in the remaining blood to maintain the heart and brain—the two organs that depend on a constant supply of oxygen to survive.

Water conducts heat much more rapidly than air and is colder than the mammalian body temperature of about 37 °C (98.6 °F). Cetacean evolution has countered this problem in three ways: reducing external appendages that lose heat, developing an insulating layer of blubber, and developing countercurrent circulation to minimize heat loss. The reduction of various appendages as mentioned above also facilitates locomotion in water.

In whales, a layer of the skin (dermis) has evolved into a blanket of blubber, which is extremely rich in fats and oils and therefore conducts heat poorly. This blanket covers the entire body and is up to 30 cm (12 inches) thick in large whales, making up a significant portion of the animal's weight. The oil yield of blubber from a blue whale, for example, was up to 50 tons.

The most important mechanism in cetacean thermoregulation is the development of countercurrent blood exchange, an adaptation that allows the animal to either conserve or dissipate heat as needed. Blood that drains from the surface of the skin has been cooled by close contact with the external environment, and it can return to the cetacean's heart via two different routes. If it returns by the peripheral route, the blood courses back to the heart through superficial veins, where it continues to lose heat and arrives at the heart cool. This dumps the animal's excess heat to the environment. Such heat shedding is particularly important to large whales because of their enormous surface area-to-volume ratio. If, however, the body temperature of the whale is already cool, the oxygen-depleted venous blood can instead return to the heart through vessels that are wrapped around arteries carrying warm blood to the periphery of the animal. Along this route the venous blood is warmed by the arterial blood and arrives at the heart warm. The arterial blood, having transferred its heat into the venous blood rather than the environment, arrives precooled at the surface of the skin.

Feeding Adaptations

Killer whale (Orcinus orca).

Before cetaceans evolved aquatic adaptations, they had a fully differentiated set of teeth (heterodont dentition), including incisors, canines, premolars, and molars. As the animals became more adapted to aquatic locomotion and lost the ability to manipulate food with their forelimbs, they started grabbing their food and swallowing it whole. In toothed whales (suborder Odontoceti), heterodont dentition declined and was replaced with a homodont dentition in which every tooth is a simple cone. The number of teeth varies among toothed whales, from two in the beaked whales (family Ziphiidae [Hyperoodontidae in some classifications]) to 242 in the La Plata river dolphin (Pontoporia blainvillei), to allow efficient capture of prey. Baleen whales (suborder Mysticeti), on the other hand, have lost all teeth in both jaws and instead have two rows of baleen plates in their upper jaws only. This apparatus enables baleen whales to consume vast quantities of small prey in a single mouthful.

In general, whales have relatively large mouths. The mouth of one adult bowhead, or Greenland right whale (Balaena mysticetus), measures five metres long and three metres wide and is the biggest oral cavity on record. The stomach in cetaceans is composed of four compartments: forestomach, main stomach, connecting chambers, and pyloric stomach. The forestomach is actually a dilation of the esophagus and is lined with simple epithelium (layers of flattened cells). It acts merely as a holding chamber and therefore is not a true stomach. The main stomach, lined with active gastric epithelium, is the first true digestive compartment, and it is followed by the small connecting chambers and the pyloric stomach. From there, food enters the small intestine through the pyloric sphincter and the duodenal ampulla. Most cetaceans do not have a cecum or appendix, and in most there is no anatomic difference between the small and the large intestine.

Greenland right whale, or bowhead (Balaena mysticetus).

Senses

The sensory system of any animal can be divided into somesthetic senses—those relating to the whole body—and special senses associated with particular organs such as the eyes and ears. Somesthetic senses are broken down into exteroceptive (initiated by stimuli outside the body), proprioceptive (initiated within the body, determining the orientation of body parts relative to one another and the orientation of the body in space), and visceral (usually from internal organs and usually painful). Cetaceans, as far as is known, are subject to the familiar exteroceptive sensations. For example, captive and stranded animals respond to stimuli of touch, pain, and heat. Because precise assessment of the other somesthetic modalities (proprioceptive and visceral) is difficult, scientists have simply assumed their presence.

The special senses respond to stimuli registered by specialized organs or tissues. One way to quantify the presence of a special sense in an animal is to consider the organs involved.

Smell

The sense of smell can be defined as those sensations carried from nose to brain by the olfactory nerve. Toothed whales have lost the olfactory nerve, so by definition they are incapable of smelling.

On the other hand, they do use "quasi-olfaction". Baleen whales have retained this nerve and have a reduced area for olfaction in the nasal passage, but this sense is active only while the animal is breathing at the surface.

Taste

Captive dolphins (family Delphinidae) commonly exercise food taste discrimination that is comparable to the human ability, in spite of the fact that the presence of taste buds in cetaceans has not been demonstrated. Regardless, dolphins have been shown to be sensitive to the standard four qualities of taste: sweet, salty, sour, and bitter. It has been established that the bottlenose dolphin (Tursiops truncatus) has a highly effective sense, called quasi-olfaction, operating through pits in the back of the tongue. This sense permits dolphins to experience what would be classified as smell, but quasi-olfaction does not involve the nasal passages.

Sight

Cetaceans have well-developed eyes and good vision. The popular notion that whales have reduced vision is probably based on the relative size of their eyes, but this assumption is functionally incorrect. Vision in both the water and the air has been experimentally evaluated in captive dolphins and found to be excellent. They have binocular vision over at least part of the visual field but are largely insensitive to colour. In one genus of river dolphin (Platanista of the muddy Ganges and Indus rivers), the eyes are reduced to organs that can detect only the difference between light and dark. The external opening for the eye is a slit only 2−3 cm (about an inch) long.

Hearing

Whales and dolphins have long been known to possess an acute sense of hearing. When approaching whales, whalers muffled their oars to prevent the animals from hearing them. Research done with captive animals in the 1950s quantitatively demonstrated that dolphins both produce and are sensitive to sounds into the ultrasonic range. Dolphins and porpoises were found to have the ability to derive information about their environment by listening to echoes of sounds that they have produced (echolocation). The amount of information obtained by an echolocating dolphin is similar to that obtained with the eyes of a sighted human.

The sound sensitivity of dolphins falls off near the bottom of the human acoustic spectrum (40–50 hertz), but this is the beginning of the range used by the large baleen whales. Fin whales (Balaenoptera physalus) and blue whales have been recorded producing subsonic sounds around 10 hertz and are capable of producing extremely loud noises at those frequencies. The strength of these vocalizations enabled one blue whale to be followed by fixed hydrophone arrays on the ocean bottom for 43 days over a course of 2,700 km (1,700 miles).

Magnetic Sensitivity

Much interest has been shown in various animals' ability to sense the Earth's magnetic field. It has been demonstrated that birds and fish use magnetoreception in migration, and theories to explain why cetaceans beach themselves in mass strandings have included magnetic detection. Although magnetite has been found in some skulls of the common dolphin (Delphinus delphis), it has not been found in other specimens of the same species, and no conclusive data indicate its biological use.

Distribution and Migration

Cetaceans are distributed in all the world's oceans from the far polar reaches to the Equator. They concentrate in areas of increased biological productivity, such as upwellings where there is an abundant supply of food. Some species are coastal, and some are pelagic, dwelling farther offshore. The centres of the ocean basins appear not to have any concentrations of whales or dolphins. Some small cetaceans are distributed in major river systems, particularly river dolphins of the family Platanistidae. Members of this family are found in the Amazon, Orinoco, La Plata, Yangtze, Ganges, and Indus rivers and surrounding drainage waters. Members of other families, particularly the Delphinidae and Phocoenidae, spend part of their time in fresh water.

Humpback whale (Megaptera novaeangliae).

As a rule, large whales have north and south seasonal migrations, spending summers in high latitudes near the poles, where there is an abundant food supply, and moving toward the Equator in the fall to breed. Some populations of these species, however, reside in one locality all year. One of the greatest migrations is undertaken by the California population of the gray whale, which summers in the Bering and Chukchi seas of the Arctic and winters in lagoons off the coast of Baja California—a journey of 5,000 km (about 3,000 miles) each way.

Migratory whales usually do not cross the Equator, and that has led to the development of genetically separate populations in the north and south ocean basins. However, one humpback whale (Megaptera novaeangliae) was photographically identified near the Antarctic Peninsula and was later sighted on the coast of Colombia, having covered at least 8,334 km in both the South Atlantic and North Atlantic ocean basins. Some sperm whales (Physeter catodon) sexually segregate on their migrations, with larger and older males going much farther in a polar direction during the summer.

Diet

All cetaceans are carnivores and do not consume plants or algae as food. The large baleen whales eat schooling organisms that range in length from minute drifting mollusks, copepods (1 cm or less), krill (1–5 cm), and small fish and squid up to about 40 cm. All these are consumed by whales in vast quantities with each concentrated mouthful. Certain smaller baleen whales, such as the minke whale (Balaenoptera acutorostrata), also pursue individual fish up to 1 metre long.

Toothed whales, which range in length from about 1 metre for the finless porpoises (Neophocoena phocaenoides and N. asiaeorientalis) to 20 metres for the sperm whale, eat an enormous variety of prey ranging from small shrimp, fish, and squid to bluefin tuna (3 metres long) and giant squid.

Cetaceans void their solid digestive waste products as pastelike feces, enabling the retention of intestinal water. Liquids are excreted in the urine. Biologists do not entirely understand the function of ambergris, which seems to be a normal digestive secretion of the sperm whale.

Reproduction

Cetacean breeding is seasonal, usually in the winter, and females normally calve once every two years. As mammals, they reproduce by internal fertilization. The testes and penis of the male are internal, but the penis is capable of being extended and introduced to the female during mating. After the female's egg has been fertilized, she carries the fetus for about a year, although some toothed whales have gestations of up to 18 months. Cetaceans give birth tail first, opposite of most terrestrial mammals. The mothers produce extremely rich milk for their young—50 percent fat is common. The mammary glands are paired and located at the lower abdomen, just forward of the anal-genital slits. One calf is born (multiple fetuses have been found, but never live twins), which is weaned at six months to a year but continues to grow rapidly until 5 to 10 years of age. The largest whale, the blue whale, is born with a length of about 7.3 metres and a weight of about 3 tons; it grows an average of 0.3 metre per week and gains weight at a rate of 90 kg (nearly 200 pounds) per day.

Bottlenose dolphin.

Age and growth

Generally, male cetaceans reach a slightly greater size than females, but there are many exceptions; female baleen whales and some of the beaked whales, river dolphins, and porpoises tend to be slightly larger than males of their species. Male sperm whales, on the other hand, are on average 50 percent larger than females. Physical maturity is defined specifically in cetaceans as the point when all of the vertebrae stop growing, which occurs at an age of about 8–25 years. Historical estimates of whale longevity are limited because of difficulty measuring age in older whales. Modern techniques have determined some fin whales to be 100 years old, some humpbacks 96, and some blue whales 90. But the longest-lived cetacean by far is the bowhead, a right whale that can survive for more than 200 years.

Scientists examining the remains of a 70-foot-long female blue whale
(Balaenoptera musculus) on the rocky shore near Fort Bragg, California.

Sexual maturity occurs at an age of about 6–10 years and is defined in cetaceans as the age at which the females start ovulating and are capable of becoming pregnant. In most species, females continue to ovulate throughout their lives and thus can produce young until death. However, females in several toothed whale species—including killer whales, false killer whales, short-finned pilot whales, belugas, and narwhals—experience menopause (in which ovulation stops), and the rest of their lives are spent in a post-reproductive state. Authorities speculate that this adaptation may have evolved in order for older females to assist with the rearing of young.

Abundance

Northern right whale (Eubalaena glacialis).

Counting animals that can be spread over a wide areas of the world's oceans and are visible for only a few seconds while they breathe is extremely difficult and expensive. The largest populations existed in the oceans around Antarctica, where harsh and remote conditions make biological research difficult and infrequent. The abundance of cetaceans is thus hard to estimate accurately, but whale populations have varied over the years, depending largely upon human activities.

Japanese factory ship hauling a minke whale through
a slipway in the ship's stern, 1992.

Biologists estimate that there were 228,000 blue whales and 548,000 fin whales in the world's oceans when modern whaling began in the early 20th century. At the beginning of the 21st century, there were an estimated 14,000 blue whales and 120,000 fin whales left. California gray whales were thought to number 20,000 in 1847, then were hunted until they were thought to be extinct in the 1920s. Since then the species has recovered under protective legislation, and its population has been estimated to be more than 26,000.

Until the early 21st century, the only cetacean population to be completely exterminated was the Atlantic gray whale, which was gone in the early 1700s; however, the baiji, or Chinese river dolphin (Lipotes vexillifer), a species restricted to the Yangtze River (Chang Jiang), was widely believed to be extinct. In addition, some smaller cetacean species with limited distributions—such as the Gulf of California porpoise (Phocoena sinus), Indus susu (Platanista minor), and Ganges susu (Platanista gangetica)—could be in immediate danger of becoming extinct. Furthermore, many other cetacean species, such as the gray whale and the northern right whale (Eubalaena glacialis), continue to be menaced by ship collisions, pollution, entanglement in commercial fishing equipment, and illegal hunting.

Ganges river dolphin, or susu (Platanista gangetica).

Diseases and Parasitism

Cetaceans can suffer from many of the same diseases and parasites that afflict humans and other mammals: cancer, arthritis, pneumonia, lungworms, tapeworms, and roundworms, to name just a few. In the 1980s various dolphin species experienced epidemics of a morbillivirus, a disease similar to distemper and measles.

Of the parasites and commensal organisms, some are also found on fish and marine turtles, and others are specific to cetaceans. Commensal barnacles are most visible on humpbacks and gray whales, although they occur to a lesser extent on many other baleen and toothed whales. Xenobalanus globicipitis, a unique type of small pseudo-stalked barnacle, occurs on the appendages of cetaceans, including the common bottlenose dolphin. Stalked barnacles can also occur on exposed teeth and can be particularly striking on the tusks of beaked whales.

Different types of commensal or parasitic crustaceans inhabit whales. There is a small commensal copepod, Balaenophilus, that eats the algae on the baleen of some rorqual species. A specific family of amphipods (Cyamidae) called whale lice routinely infest right, humpback, and gray whales and also occur opportunistically on most species of baleen and toothed whales, particularly around wounds. They appear to eat sloughed skin. Another crustacean is clearly parasitic and is a member of the caligoid copepod genus Pennella. It is commonly about 2–10 cm long and lives with its body buried in the blubber or skin of cetaceans and fishes.

Minke whale (Balaenoptera acutorostrata).

Small (4–7 cm) circular scars left in the skin of whales, dolphins, and fish were a mystery during the early 20th century. Explorers reported that the scars were created by an unknown organism that they called the "DWB" or "demon whale biter," which cleanly removed hemispheric chunks of blubber as though extracting them with a razor-sharp scoop. The creature responsible was finally identified in the 1950s as a grazing predator, the cookie-cutter, or cigar, shark (genus Isistius).

Locomotion

Swimming

Cetaceans swim by using vertical tail movements that drive the horizontal flukes up and down, powered by the long epaxial and hypaxial muscles that lie along the spine. The tail flexes through a point between the dorsal fin and the anus, while the thorax and abdomen are relatively inflexible. The body itself acts like a spring to propel the animal through the water with minimal energy.

Sperm whale (Physeter catodon).

Much was written about the speeds of cetaceans in the mid-20th century. It seemed that cetaceans could exceed the speed at which turbulence would make locomotion energetically very expensive. However, the swimming-speed figures were estimates that turned out to be very high. Further investigation found that, regardless of size, the cruising speed of most cetaceans is about 2 metres per second (about 7 km, or 4 miles, per hour). A combination of biomechanical and hydrodynamic factors make this an efficient speed at which to travel. Maximum speeds, however, vary greatly between species.

Common dolphins (genus Delphinus) have been observed keeping pace with boats for a considerable period of time at 36 km/hr (kilometres per hour). Researchers trained Pacific bottlenose dolphins (genus Tursiops) to swim in an open-water environment, thus removing the spatial limitations of a pool while conserving experimental controls. They found that the dolphins could sprint at 29.9 km/hr for 7.5 seconds and could maintain speed at 21.9 km/hr for 50 seconds. When dolphins ride a bow wave, they coast at the speed of the ship while expending very little energy. Fin and blue whales can swim fast enough that a boat must travel in excess of 30 km/hr to catch up to them, and they can maintain speeds of 33–37 km/hr for periods of up to 10–15 minutes. Sonar records indicate that fin whales can sprint at 48 km/hr. Right, humpback, and gray whales, however, can seldom swim faster than 9 km/hr. Sperm whales can cruise at 7.5 km/hr and swim up to 36 km/hr in spurts. The fastest cetacean appears to be the sei whale (Balaenoptera borealis), recorded moving at speeds up to 65 km/hr along the ocean surface.

Dolphins (family Delphinidae) and river dolphin (family Platanistidae).

Breathing and Diving

Cetaceans surface periodically to breathe, and the intervals between breaths vary depending on what the animal is doing. Intervals may range from about 20 seconds for dolphins that are actively swimming to 5–10 minutes for a resting blue whale. A common breathing pattern in large whales is to breathe every 20 seconds for 8–10 breaths and then dive for about 10–15 minutes. Most whales stay in the upper 100 metres of water. Deep-diving whales—such as the sperm whale, which has been recorded diving to depths of 1 km—may stay down for an hour. The longest recorded dive is that of a harpooned bottlenose whale (Hyperoodon ampullatus) that dived for two hours, surfaced, and then dived again. Patterns of locomotion and breathing are very important to whale watchers identifying whales at a distance, as different species show different blow heights and shapes. Right whales, for instance, have an unequal inclination to their two nasal passages, so their blows appear in pairs. Humpbacks and gray whales have blows that appear low and wide (bushy), and sperm whales have a bushy blow that is angled to the left and forward.

Leaping and Wave Riding

Small cetaceans "porpoise" when they are swimming rapidly; that is, they rise out of the water in a low leap that keeps the head clear of the water for breathing. Spinner dolphins (Stenella longirostris) frequently leap out of the water while spinning on their long axis, hence their common name. Trained porpoises and dolphins can leap straight up as high as six metres. Leaping is very rare in large whales, but some rorquals (genus Balaenoptera) have been photographed jumping clear of the water.

Whales such as humpbacks (Megaptera novaeangliae) communicate by producing
low-frequency sound waves. The animals move sufficiently far beneath the ocean
surface before vocalizing, which enables their signals to be heard over hundreds of kilometres.

Many small cetaceans play around moving boats, where they bow-ride, taking advantage of their ability to bodysurf and essentially enjoying the free ride in the bow wave created by the vessel. They also practice this behavior around large whales that are swimming fast enough to produce a bow wave.

Behavior

Social Behavior

All cetaceans are social to some extent. The minimum group of mother and calf is commonly expanded to a nuclear family or a group of closely related individuals. A group of cetaceans that normally feed and travel together is called by various names: school, herd, pod, or gam. It is often difficult to define or measure, as its members can be spread over kilometres of ocean but still be in contact with one another. Sometimes these schools coalesce into even bigger groups of more than 1,000. Groups of whales can persist for many years, and studies of coastal dolphins have shown long-term association of dolphins with their mothers. Groups (particularly of small toothed whales) frequently associate with other cetacean species. For example, associations between pilot whales and bottlenose dolphins have been observed, as have associations between common dolphins and fin whales.

Fin whale (Balaenoptera physalus).

Play is a common behaviour, especially among young animals. Play allows individuals to practice and perfect behaviour patterns, such as aggression, that will be socially useful later in life; a significant portion of play is sexually oriented. Captive dolphins have also been observed playing with fish, birds, and turtles.

Many cetaceans exhibit epimeletic behaviour, in which healthy animals take care of another animal that has become temporarily incapacitated. This is evident when a wounded or sick whale is supported by others or in cases when a dolphin (usually the mother) pushes a dead calf around.

Cetaceans show fright by fleeing from a situation or by bunching up and "milling." The former response has been utilized by fishermen, who drive a whale or school of dolphins into a situation where they can kill it. Milling has been seen in dolphin schools driven into an enclosure or caught in a net; the animals move in a circle or eddying mass, and at the height of this reaction they stop swimming, sink, and die.

Aggression and Defense

Aggression is common among cetaceans and is seen in normal herd behaviour and feeding. One form of aggression helps to establish social hierarchy: the dominant animal nips the less-dominant animal, which produces the tooth scars seen on every adult in the dolphin family (Delphinidae). Mating behaviour also involves biting, as one of the ways males compete for females is by biting and raking the teeth over another male. Adult male beaked whales (family Ziphiidae) have very densely ossified rostra (beaks) used as weapons in combat for females. Another more dangerous means of aggression is head butting. Cetaceans can ram their heads into other individuals and kill them. This has been seen in captivity and in aggressive behaviours toward other species such as sharks and accounts for many of the broken ribs and vertebrae seen in stranded animals.

Baird's beaked, or giant bottlenose, whale (Berardius bairdii).

Normally, aggression is associated with members of the same species or as a defense response to predation from other species. Although cetaceans can defend themselves by utilizing the behaviours of intraspecific aggression (biting, ramming, and butting), the primary weapon that cetaceans have for self-defense is the tail. Cornered whales slash sideways with their flukes and can incapacitate a bigger whale or a boat. Head butting as a form of defense was immortalized in the 1820 sinking of the whaling ship Essex by a sperm whale.

Courtship and Mating

Sexual behaviour starts early in cetaceans. Young dolphins engage in exploratory sexual behaviour involving their mothers and other members of the school. Self-stimulation is common in both sexes. Male cetaceans perhaps use their penises as a manipulation organ in much the same way that people use their hands. This exploratory behaviour gradually becomes courtship and mating behaviour.

Courtship involves physical and acoustic displays, such as the elaborate songs of male humpbacks, and leads to contact with the flippers and other parts of the body. Successful courtship culminates in mating. Copulation is relatively brief in cetaceans. It can be secretive, or it can be boisterous as in the mating displays of right and gray whales, when a number of males attempt to mate with a single female.

Feeding

Cetaceans hunt as individuals or in schools. When hunting in schools, dolphins or whales herd their prey in order to concentrate a large volume before eating. Hunting alone is preferred where prey is more scattered.

Killer whale (Orcinus orca).

Before they swallow their food, toothed whales disable it; biologists think that some can stun their prey by emitting a high-energy burst of sound. Normally, cetaceans eat animals that can be swallowed intact, as their teeth are shaped for holding, not chewing. If, however, the prey is too large to swallow in one bite, it is ripped into chunks. Killer whales (Orca orcinus) have been seen to grab seals and shake them in the air so hard that the bodies come apart.

Orca, or killer whale (Orcinus orca).

Baleen whales also herd their prey like toothed whales, but they engulf it in either of two feeding methods: gulp or skim. In gulp feeding, the whale opens its mouth to take in a huge mouthful of water, closes its mouth, strains the water out through the baleen apparatus along the sides of the mouth, and swallows its prey. Gulp feeding is common in rorquals, which have ventral grooves that stretch to enlarge the oral cavity. One of the rorquals, the sei whale, as well as the nonrorqual baleen whales (right, bowhead, pygmy right, and gray), skim-feed by locating a concentration of zooplankton prey and swimming through it with the mouth open. Skimming may last up to several minutes until the whales close their mouths to swallow what they have filtered from the water.

Sei whale (Balaenoptera borealis).

Sleep

Breathing is a conscious activity in cetaceans; they must consciously breathe, or they will drown. Therefore, they cannot enter into what humans understand as unconscious sleep; instead, they have periods of little activity but not total inactivity. Studies of dolphins have revealed that they shut down half of their brain during sleep. The other half of the brain stays awake to signal when to rise to the surface to breathe and to watch for predators and obstacles. Large whales appear to surface-sleep. Floating horizontally just below the water's surface, they move their flukes periodically to rise above the water for a breath.

Intelligence

Although several cetaceans are easily trained and much has been theorized about the possible intelligence of whales and dolphins, little is known for certain. Some researchers equate brain size with intelligence, reasoning that cetaceans should have the capacity for intelligence because they have relatively large brains. The human brain averages about 1.2 kg, the bottlenose dolphin brain about 1.8 kg. The largest cetacean brain recorded was a sperm whale's, weighing 9.2 kg. However, cetaceans may use their increased brain weight for processing acoustic information. In any event, it seems unproductive to compare species with which it is difficult even to communicate until a definition of nonhuman intelligence has been refined.

Stranding

Stranding is a phenomenon that has long fascinated people, and there is fossil evidence of mass strandings from before humans evolved. Many stranded cetaceans are found already dead, and it is not known if they were alive and conscious when they stranded themselves. When a whale or dolphin dies offshore, it usually sinks; if the water is shallow enough to permit decomposition gases to form, it will float ashore, so some stranding represents normal mortality. If infection or some other factor interferes with a cetacean's ability to navigate, it could come ashore while still alive—though most cetaceans have difficulty out of water and usually die. These cases are known (alive or dead) as single strandings. Sometimes up to several hundred toothed whales swim ashore, and this phenomenon is known as a mass stranding.

There are no records of the mass stranding of baleen whales; all such events have involved only toothed whales that normally live offshore and may not be familiar with physical borders. Perhaps not realizing that the ocean has a bottom and sides, they may somehow enter shallow water and find themselves unable to deal with the strange environment. Because they are also members of extremely social species and may be kept together by group ties, they may have even greater difficulty extricating themselves. In any case, biologists are beginning to realize that cetaceans are behaviorally complex enough that a simple blanket explanation of mass stranding is not likely to be valid. Biologists have tried to attribute mass stranding to a number of causes:

1. Something wrong with the leader of a group.

2. Epidemic disease.

3. Getting lost in pursuit of prey.

4. Parasitic infestation that affects the hearing.

5. Following migratory routes laid down by remote ancestors.

6. Magnetic anomalies that lead the school astray.

7. Behavioral reversion to a period when cetacean ancestors were terrestrial and land was a haven.

8. Fright reaction to predators.

9. Failure of echolocation signals to work properly in shallow water.

10. Overpopulation.

11. Suicide.

Sound Production and Communication

All cetaceans produce sound, some more extensively than others, and they primarily use the larynx for this purpose. At one time it was argued that the cetacean larynx was incapable of generating sound because it does not have vocal cords. However, vocal cords are restricted mainly to primates; dogs and cats, for example, have vocal folds, and both baleen and toothed whales possess structures that modify sound. Baleen whales have laryngeal pouches, and toothed whales have accessory air sacs and fat bodies in their noses. In addition, toothed whales can generate high-frequency sounds in their nasal passages.

Cetacean sounds can be roughly divided into communication signals and echolocation signals. Communication does not necessarily imply language, and it can simply be one-way, as when one dolphin knows another is present because the second dolphin is vocalizing. Echolocation, which involves generating certain sounds and listening to the echoes of those sounds, has been recognized in toothed whales but not baleen whales. Toothed whales use extremely high frequencies, on the order of 150 kilohertz, for refining spatial resolution from their echoes. They are capable of "seeing" into and through most soft objects such as other dolphins, though the effectiveness of toothed whale echolocation drops off at distances greater than about 100 metres. To produce such high frequencies, toothed whales possess modified tissues associated with the blowhole on the right side of the head; the left side is not modified, and the result is skull asymmetry. This condition is extreme in sperm whales, which is not surprising, as most of the head is involved with sound production. The head of a 16-metre adult male sperm whale is about 6 metres long, 3 metres high, and 3 metres wide, a mass of tissue that can weigh about 20 tons. The bulk of it is occupied by the spermaceti organ and a fatty (adipose) cushion, both of which somehow function in the emission of sound for echolocation and were known by whalers as the "case" and the "junk," respectively. The junk of the sperm whale is the fatty structure found in the forehead of other toothed whales and known by whalers as the "melon" because of its pale yellow colour and uniform consistency. Baleen whales generate sounds at frequencies that are audible to humans (sonic) or below that range (subsonic). Some of their vocalizations are very loud; biologists have recorded extremely low sounds (12.5–200 hertz) from a blue whale and have claimed they are the loudest sounds known from any animal. The songs that humpbacks use for courtship were brought to public awareness in 1971. Baleen whales (mysticetes) use calls like these for communication and possibly for low-frequency long-range echolocation in orientation and navigation. Their low-frequency sounds are powerful enough that mysticetes might be able to communicate across entire ocean basins.

PHYSIOLOGY OF CETACEANS

Circulation

Cetaceans have powerful hearts. Blood oxygen is distributed effectively throughout the body. They are warm-blooded, i.e., they hold a nearly constant body temperature.

Respiration

Cetaceans have lungs, meaning they breathe air. An individual can last without a breath from a few minutes to over two hours depending on the species. Cetacea are deliberate breathers who must be awake to inhale and exhale. When stale air, warmed from the lungs, is exhaled, it condenses as it meets colder external air. As with a terrestrial mammal breathing out on a cold day, a small cloud of 'steam' appears. This is called the 'spout' and varies across species in shape, angle and height. Species can be identified at a distance using this characteristic.

The structure of the respiratory and circulatory systems is of particular importance for the life of marine mammals. The oxygen balance is effective. Each breath can replace up to 90% of the total lung volume. For land mammals, in comparison, this value is usually about 15%. During inhalation, about twice as much oxygen is absorbed by the lung tissue as in a land mammal. As with all mammals, the oxygen is stored in the blood and the lungs, but in cetaceans, it is also stored in various tissues, mainly in the muscles. The muscle pigment, myoglobin, provides an effective bond. This additional oxygen storage is vital for deep diving, since beyond a depth around 100 m (330 ft), the lung tissue is almost completely compressed by the water pressure.

Organs

The stomach consists of three chambers. The first region is formed by a loose gland and a muscular forestomach (missing in beaked whales), which is then followed by the main stomach and the pylorus. Both are equipped with glands to help digestion. A bowel adjoins the stomachs, whose individual sections can only be distinguished histologically. The liver is large and separate from the gall bladder.

The kidneys are long and flattened. The salt concentration in cetacean blood is lower than that in seawater, requiring kidneys to excrete salt. This allows the animals to drink seawater.

Senses

Cetacean eyes are set on the sides rather than the front of the head. This means only species with pointed 'beaks' (such as dolphins) have good binocular vision forward and downward. Tear glands secrete greasy tears, which protect the eyes from the salt in the water. The lens is almost spherical, which is most efficient at focusing the minimal light that reaches deep water. Cetaceans are known to possess excellent hearing.

At least one species, the tucuxi or Guiana dolphin, is able to use electroreception to sense prey.

Ears

The external ear has lost the pinna (visible ear), but still retains a narrow external auditory meatus. To register sounds, instead, the posterior part of the mandible has a thin lateral wall (the pan bone) fronting a concavity that houses a fat pad. The pad passes anteriorly into the greatly enlarged mandibular foramen to reach in under the teeth and posteriorly to reach the thin lateral wall of the ectotympanic. The ectotympanic offers a reduced attachment area for the tympanic membrane. The connection between this auditory complex and the rest of the skull is reduced—to a single, small cartilage in oceanic dolphins.

In odontocetes, the complex is surrounded by spongy tissue filled with air spaces, while in mysticetes, it is integrated into the skull as with land mammals. In odontocetes, the tympanic membrane (or ligament) has the shape of a folded-in umbrella that stretches from the ectotympanic ring and narrows off to the malleus (quite unlike the flat, circular membrane found in land mammals.) In mysticetes, it also forms a large protrusion (known as the "glove finger"), which stretches into the external meatus and the stapes are larger than in odontocetes. In some small sperm whales, the malleus is fused with the ectotympanic.

The ear ossicles are pachyosteosclerotic (dense and compact) and differently shaped from land mammals (other aquatic mammals, such as sirenians and earless seals, have also lost their pinnae). T semicircular canals are much smaller relative to body size than in other mammals.

The auditory bulla is separated from the skull and composed of two compact and dense bones (the periotic and tympanic) referred to as the tympanoperiotic complex. This complex is located in a cavity in the middle ear, which, in the Mysticeti, is divided by a bony projection and compressed between the exoccipital and squamosal, but in the odontoceti, is large and completely surrounds the bulla (hence called "peribullar"), which is, therefore, not connected to the skull except in physeterids. In the Odontoceti, the cavity is filled with a dense foam in which the bulla hangs suspended in five or more sets of ligaments. The pterygoid and peribullar sinuses that form the cavity tend to be more developed in shallow water and riverine species than in pelagic Mysticeti. In Odontoceti, the composite auditory structure is thought to serve as an acoustic isolator, analogous to the lamellar construction found in the temporal bone in bats.

Cetaceans use sound to communicate, using groans, moans, whistles, clicks or the 'singing' of the humpback whale.

Echolocation

Odontoceti are generally capable of echolocation. They can discern the size, shape, surface characteristics, distance and movement of an object. They can search for, chase and catch fast-swimming prey in total darkness. Most Odontoceti can distinguish between prey and nonprey (such as humans or boats); captive Odontoceti can be trained to distinguish between, for example, balls of different sizes or shapes. Echolocation clicks also contain characteristic details unique to each animal, which may suggest that toothed whales can discern between their own click and that of others.

Mysticeti have exceptionally thin, wide basilar membranes in their cochleae without stiffening agents, making their ears adapted for processing low to infrasonic frequencies.

Chromosomes

The initial karyotype includes a set of chromosomes from 2n = 44. They have four pairs of telocentric chromosomes (whose centromeres sit at one of the telomeres), two to four pairs of subtelocentric and one or two large pairs of submetacentric chromosomes. The remaining chromosomes are metacentric—the centromere is approximately in the middle—and are rather small. Sperm whales, beaked whales and right whales converge to a reduction in the number of chromosomes to 2n = 42.

ECOLOGY OF CETACEANS

Cetacean ecology describes the relationships between cetaceans and their physical and biological environment, including their interactions with prey, predators, competitors, and commensals. Studying whales or dolphins in their natural environment is a formidable challenge. This is not only due to the logistical constraints of attempting to study highly mobile, oceanic animals that spend nearly all of their lives underwater, but also because of the political and legal constraints of working on protected species, which include most cetaceans. Early insights into cetacean ecology came largely from anecdotes and observations handed down by early whalers (Herman Melville's Moby Dick is a classic example). Although much remains unknown, technological developments over the past decade have greatly facilitated our understanding of cetacean ecology. Insights gained from individuals fitted with devices that transmit data on location and depth of dive through a satellite and to a researcher's desk, from linking movements and distribution with remotely sensed data on surface water properties of the ocean, or from sophisticated shipboard equipment designed to quantify density of prey in the water column are just three examples of technological advancements that have greatly contributed to clarifying cetacean ecology.

Cetaceans include approximately 84 species; new species continue to be described. They are a diverse group. They range in size from less than 1 m long for a newborn vaquita (Phocoena sinus), to 33 m in an adult blue whale (Balaenoptera musculus); they occupy water ranging in temperature from 2 °C to over 30 °C; they exhibit a diverse array of life history strategies. Consider the sperm whale (Physeter macrocephalus) which can remain beneath the water for over an hour and dive to depths of several thousand meters, the Ganges river dolphin (Platanista gangetica) which inhabits fresh water so turbid it is functionally blind, beaked whales of the genus Mesoplodon which are so pelagic and so elusive that some have never been seen alive in the wild, the gray whale (Eschrichtius robustus) which annually migrates some 15,000 to 20,000 km between breeding and feeding areas, and the bowhead whale (Balaena mysticetus) which uses its rostrum to break ice in the Arctic.

One of the challenges of ecology is to search for pattern within diversity. Despite their diversity of form, behavior, and habitat, all cetaceans have some key features in common that underscore the fact that they are secondary marine forms, derived from terrestrial ancestors. That they are all air-breathing, live-bearing homeotherms provides a unifying theme.

Habitat in the Ocean

Marine habitat is largely about oceanography. Terrestrial habitat is typically defined by the interaction between physical structure (as defined by physiography, and characteristics of primary producers) and meteorological factors, whereas marine habitat is almost entirely defined by hydrographic, physical, and chemical properties of water (e.g., water masses, surface currents, fronts, eddies, island wakes). Marine habitat types can be geographically fixed, when referring to the benthos or coastal ecosystems, e.g., but in most cases, marine habitat is not static in space or time. Instead, habitat types move with the water masses and surface currents that define them. Physical structure can define marine habitat, in the case of kelp forests or coral reefs, e.g., but in most cases, this type of physical structure is completely absent from marine habitats. Light attenuates more quickly and sound travels faster and farther in water as compared to air; thus marine organisms, including cetaceans, rely less on vision and more on the auditory sense than terrestrial mammals. The base of the trophic web in marine systems is formed by planktonic organisms. There is no analogous counterpart in terrestrial systems and marine animals have therefore evolved unique and specialized morphological and behavioral adaptations for taking advantage of this prey base.

On a global scale, cetaceans have invaded a large proportion of the ocean's habitats. They inhabit coastal waters up to and including the surf zone (gray whale, some populations of bottlenose dolphins Tursiops truncatus, harbor porpoise Phocoena phocoena, Commerson's dolphin Cephalorhynchus commersonii), neritic waters over continental shelves (long-beaked common dolphin Delphinus capensis, Lagenorhynchus spp., Cephalorhynchus spp., Phocoena spp.), and the most oceanic of systems (sperm whale, Fraser's dolphin Lagenodelphis hosei, beaked whales). They are found in tropical waters (pantropical spotted dolphin Stenella attenuata), temperate seas (Risso's dolphin Grampus griseus), and polar oceans, up to and within pack ice (beluga Delphinapterus leucas, bowhead whale). They utilize much of the water column, some being confined to relatively shallow depths (most dolphins and baleen whales), and others diving to thousands of meters (sperm whale, many beaked whales). And they have invaded the world's major river systems (Ganges, Indus, Amazon/Orinoco).

Cetaceans in different habitats might be expected to show differential development of adaptations which reflect selective pressures of the environments in which they function. For example, species in polar seas must conserve heat and so, bowhead whales have relatively large bodies, thick blubber layers, and short appendages. Deep-diving species (sperm and beaked whales) must conserve oxygen and might be expected to have large blood volumes, high hematocrit, and a well developed diving response. Species that forage in low light conditions (night feeders, deep divers, species living in turbid rivers) should have well-developed echolocation abilities relative to those that function in habitats with greater light levels and better visibility.

The geographic range that a single species occupies runs from cosmopolitan to extremely local. For example, the killer whale (Orcinus orca) can be found throughout the world's oceans and, with the exception of humans, is the most wide-ranging mammal on earth. At the other extreme is the vaquita, a tiny porpoise that occupies a few hundred square km in the northern Gulf of California. And some species migrate between widely separate breeding and feeding areas; this pattern is characteristic of most (if not all) of the baleen whales.

Species–habitat Relationships

The relationship between a species and its habitat is a defining feature of its ecology. Species–habitat relationships form the basis for defining a species 'ecological niche, in turn, a driving factor in determining competitive relationships, and a species' role in communities. Species–habitat relationships identify core requirements, critical knowledge for effective management and conservation. And a solid understanding of species–habitat relationships allows for prediction of distribution and abundance, prediction that can facilitate mitigation of anthropogenic impacts such as ocean noise and climate change. Because quantification of species–habitat relationships by definition involves integration of two very different types of data: species and habitat, and because marine habitat is so often defined by oceanographic features, the study of species–habitat relationships is not straight forward. Simple correlations between a species and one or two directly measured oceanographic variables form the basis for early understanding of these relationships. More recently, two types of analytical tools have been used. Descriptive methods include overlays of species data on maps of oceanographic measures, correlation analysis, goodness of fit metrics, analysis of variance, and ordination. Modeling techniques are more sophisticated analytically, requiring parameter estimation, model selection, uncertainty estimation, and model evaluation. Many of these methods are still being developed.

In many geographic regions and for many species, we are beginning to identify those features that correlate with centers of distribution, thereby possibly identifying what may be called critical habitat. For example, some species associate with ice edges (beluga), some with continental shelf edges or seamounts (beaked whales), and some with shorelines (gray whale, bottlenose dolphin, harbor porpoise). For oceanic species habitat preferences are often defined by less obvious features: physical and chemical characteristics of the water itself, which define water masses and current boundaries. For example, some species associate with cold-water currents (Heaviside's Cephalorhynchus heavisidii, Commerson's, Peale's Lagenorhynchus australis dolphins). Blue whales in the eastern Pacific are found in relatively cool, upwelling-modified waters with high primary and secondary productivity. And in the eastern tropical Pacific, pantropical spotted dolphins and spinner (Stenella longirostris) dolphins segregate from common dolphins (Delphinus spp.) according to thermocline depth and strength, sigma-t (a measure of seawater density computed from surface temperature and salinity), and surface water chlorophyll content. These differences are statistically significant and these species-specific distribution patterns track oceanographic variation on a seasonal and interannual basis.

Prey are likely the drivers of these species–habitat relationships, not the physical variables typically used in these types of analyses. In fact, there are a number of studies which have linked general distribution and movement patterns of cetacean species (humpback Megaptera novaeangliae, fin Balaenoptera physalus, long-finned pilot whales Globicephala melaena, Atlantic white-sided Lagenorhynchus acutus, bottlenose, common dolphins) with those of their prey.

Food, Feeding and Foraging

Cetacean Prey

Most of what is known about the food of cetaceans comes from data collected from dead animals, through directed fisheries, incidental mortality, or strandings. Prey of cetaceans fall into four general categories. The first prey type consists of small individuals which school at relatively shallow depths (surface to several hundred meters). These are primarily planktonic crustaceans (euphausiids, copepods, amphipods), and small fish [e.g., herring (Clupea spp.), sardine (Sardinops spp.), anchovy (Engraulidae), sandlance (Ammodytidae)]. They tend to occur in temperate or polar seas or in those tropical latitudes that are associated with high productivity. They generally occur at low trophic levels, have small body sizes, and occur in dense aggregations. Accordingly, the cetaceans feeding upon them capture multiple individuals simultaneously, have large body sizes, and have evolved filtering mechanisms (baleen) to strain prey items from the water. All mysticetes feed on this prey type.

The second prey type is comprised of larger organisms that also school at relatively shallow depths (surface to several hundred meters), or migrate up to shallow depths during the night. This includes many pelagic fi shes [e.g., hake (Merluccius spp.), pollock (Pollachius spp., Theragra spp.), myctophids (Myctophidae)] and schooling squids (Loligo spp., Dosidicus spp.) that occur throughout the world's oceans. Because these prey are larger, they generally occupy higher trophic levels and are captured individually. Their cetacean predators typically have smaller body sizes. They include all of the large-schooling dolphins (e.g., dusky Lagenorhynchus obscurus, common, striped Stenella coeruleoalba, spotted dolphins) and some small-schooling or solitary species (e.g., bottlenose, Commerson's, river dolphins Platanista spp.). These cetaceans tend to have a high tooth count, pointed teeth, and pointed snouts, all adaptations for pursuing fast, individual prey.

The third prey type is comprised of large, solitary squid (e.g., Gonatus spp.). These are most often found in deep waters throughout the world's oceans. Because of their size and solitary habits, they are captured individually. Cetacean predators of these prey include the sperm whale, dwarf, and pygmy sperm whales (Kogia spp.), all of the beaked whales (Ziphiidae), and pilot whales (Globicephala spp.). They are deep divers and tend to have reduced dentition, rounded heads, and well-developed melons, the latter perhaps indicative of the importance of echolocation for prey detection in the dark depths.

The final prey type includes species at high trophic levels that are themselves top predators. These include predatory fishes [e.g., tunas (Scombridae), sharks, salmonids], marine birds, pinnipeds, and cetaceans, including the largest of whales [rorquals (Balaenoptera spp.) and sperm whales]. Few cetaceans are able to take these prey items. They include the killer whale and, possibly, false killer whale (Pseudorca crassidens), and pilot whales. Two distinct forms of killer whales occur in waters off the west coast of North America: those that take fish and those that take mammals and birds. There is some indication that multiple ecotypes are found in Antarctic waters, and perhaps throughout the world's oceans.

Capturing Prey

Cetaceans have two main types of feeding apparatus: baleen and teeth. Baleen is used for straining prey items from the water or, in the case of the benthic-feeding gray whale, from the sediment. Teeth are used for catching individual prey items. Species with a high tooth count use them to grasp individual prey; those with a low tooth count tend to be suction feeders.

Most of what is known about prey capture strategies is relevant to cetaceans which feed on small prey that school at relatively shallow depths (the mysticetes). This is because it is relatively easy to observe these animals feeding in the wild. Mysticetes have baleen plates suspended from the roof of their mouths that they use to strain prey items from the water. The number of baleen plates, their length, and the density of baleen fibers per plate vary between species and are correlated with prey size. The Balaenidae (right whales Eubalaena spp., bowhead whales) and sei whales (Balaenoptera borealis) have the greatest number of plates with the fi nest filtering strands and feed mainly on tiny copepods. Blue whales and most other rorquals have an intermediate number of plates with coarser fi ltering strands and feed on larger prey items such as euphausiids and small fishes. Gray whales have the fewest number of plates with the coarsest strands and are largely bottom feeders, sifting benthic infauna from muddy substrate.

In addition to specializing on different prey sizes, baleen whales have specialized feeding methods that also correlate with the morphology of their baleen. "Skimmers," the right whales, swim slowly with their mouths open through dense clouds of slow-swimming copepods. "Gulpers," including most rorquals, lunge into dense schools of euphausiids or fishes with their mouths open, closing them rapidly to trap their prey. All rorquals have throat grooves that run along the ventral surface of the mouth and throat, which allow the buccal cavity to expand during a lunge, taking in huge quantities of water, and with this, prey. A variation on this type of feeding is used by humpback whales when they form "bubble nets": streams of bubbles emitted from the blowhole as the whale swims in a circular pattern toward the surface. The bubbles form an ascending curtain, which concentrates prey inside. Most of these cetaceans are solitary feeders but they regularly aggregate in areas of high prey density and, when prey are extremely dense, will feed co-operatively at times, through bubble-net feeding or in staggered echelon formations.

Cetaceans that feed on larger fish and squid that school at relatively shallow depths capture individual prey items and swallow them whole. High speed is important, as is vision. Typically these predators forage co-operatively, herding prey into tight aggregations and capturing them in turn. Acoustic signaling is presumably important for the co-ordination of schooling activities. Some cetaceans in this group feed as individuals, particularly those found in coastal areas. They show a wide range of prey capture behaviors, including slapping fish with their flukes and deliberately stranding themselves on the beach in pursuit of fishes.

Cetaceans taking large, solitary squid feed at depth, in partial to full darkness. For this reason, not much is known about how they capture prey. They probably do not feed co-operatively because their prey do not school, and because most of these cetaceans occur in small schools and are slow swimming. Most have reduced dentition, and evidence indicates that they are suction feeders, using the gular muscles and tongue in a piston-like action to suck prey into their mouths. How they are able to get close enough to their prey to suck them in remains a mystery. One intriguing idea is that they are able to partially stun prey with echolocation bursts.

Cetaceans that prey upon top predators show a wide range of prey capture methods and hunt as individuals as well as co-operatively in groups, depending upon prey size and characteristics. For example, killer whales may take pinnipeds by beaching themselves intentionally to grab adults and pups from rookeries but hunt cooperatively to take dolphins and large whales. Co-operative behaviors include prey encirclement and capture, division of labor during an attack, and sharing of prey.

Locating Prey

Most cetaceans are visual predators, at least in part. For odontocetes, echolocation is equally important in locating and targeting prey, more so than vision in some species. Although only confirmed for a handful of captive species, all odontocetes are assumed to be able to echolocate and to use this sense extensively when foraging. At present, there is no evidence that mysticetes have the ability to echolocate, although they do produce low frequency sounds that travel long distances (hundreds of km). The long wavelengths of these pulses cannot resolve features finer than the wavelengths themselves (tens of m), and so, it is doubtful that they could be used to locate and target prey patches. The effective range of vision and echolocation is a function of water clarity and the specific echolocation abilities of a species, but both are probably limited to distances on the order of hundreds of meters to a few km.

On a larger spatial scale, patchiness and variability in space and time are characteristic of most marine ecosystems and little is known about how cetaceans locate prey in such environments. Presumably, many species simply travel large distances in a continuous search. This is particularly likely to be the case in regions of low productivity, where prey patches are few and far between, such as the oceanic tropics. Here, schooling may increase the chances of encountering a patch (the more eyes and ears, the better), and dolphin schools have been observed moving through the water in wide line-abreast formations, apparently searching for prey.

There are circumstances under which prey occur predictably in space and time, and it is likely that cetaceans search for and exploit these opportunities. For example, oceanographic features (e.g., boundaries between currents, eddies, and water masses) increase prey abundance or availability by enhancing primary production, by passively carrying planktonic organisms, and by maintaining property gradients (e.g., fronts) to which prey actively respond. Topographic features also (e.g., islands, seamounts) are sites of prey aggregation. Therefore, a good foraging strategy is simply to locate these physical features and many species of cetaceans (right, blue, fin, humpback, sperm, killer whales, spinner, Risso's, common, Atlantic spotted Stenella frontalis dolphins) have been found to associate with them.

Many species of cetaceans locate and associate with predictable point sources of prey. For example, killer whales aggregate around pinniped rookeries when young seals and sea lions are weaning. Rough-toothed dolphins (Steno bredanensis) associate with flotsam in the oceanic tropics, which serves to aggregate communities of animals at a wide range of trophic levels. A wide variety of cetaceans associate with fishing operations to take their discards or their target species.

And there are times when prey are more accessible than others. The pelagic community of fishes and invertebrates, which live at depth during the day but migrate to the surface at night, provides an opportunity for cetaceans to predictably locate prey near the surface, and some dolphins (spotted, spinner, dusky, common) are known to feed on organisms in this community at night.

Cetacean Predators

By far the most important non-human predator of cetaceans is the killer whale. Their pack-hunting behavior allows them to take everything from the fastest dolphins and porpoises to the largest whales, including blue and sperm whales. Other predators known to occasionally prey on smaller or weakened individuals include large sharks, and possibly false killer and pilot whales. Polar bears (Ursus maritimus) take cetaceans along the ice edge.

The ecological significance of this predation pressure in the lives of whales and dolphins is difficult to assess, but it may be significant. Individual large whales often show signs of killer whale tooth rake marks on their flippers, fins, and flukes, and up to one-third of the bottlenose dolphins off eastern Australia bear shark bite scars, suggesting that they regularly encounter predators. It has been hypothesized that large whales may undergo their annual migrations to reach calving grounds in areas of lower killer whale densities (i.e., the tropics). Aggregative behavior is a common defensive strategy among prey species and it is possible that schooling evolved in dolphins primarily as a defense mechanism against predators. These kinds of behavioral adaptations have cascading effects influencing not only distribution and abundance, but also social structure, timing and mode of reproduction, foraging strategies, and speciation patterns. Although its significance has been downplayed, the degree to which predation (top-down forcing) has structured cetacean ecology may have been under-estimated.

Sharks are the other main predator of cetaceans. Large-bodied sharks can maim or kill individual cetaceans but are likely significant predators only for smaller bodied dolphins or perhaps tiny calves of larger bodied cetaceans. The tiny cookie-cutter shark (Isistius brasiliensis) can also be considered a cetacean predator. These tropical sharks regularly take scoops of skin and muscle from many species of cetaceans and while these wounds are not fatal, they leave scars for the remainder of an individual's life.

Schooling

Like many animals, cetaceans form aggregations for two main reasons: feeding and protection. Feeding can bring animals together in passive aggregations in areas of high resource abundance. Alternatively, animals may actively seek others to take advantage of benefits provided by other school members. Schools also serve to protect members from predation, by providing cover for individual members, by confusing predators with synchronized movements of many individuals, by reducing the probability of predation on any one individual, by increasing the chance of detection of a predator, and by providing for co-ordinated defense. Occurring in large groups also increases the potential for social interactions, including reproduction; this may only be a secondary benefit of schooling.

The majority of cetaceans occur in schools, although there are some species that regularly occur solitarily or in very small groups of pairs or trios (many mysticetes, large male sperm whales, most beaked whales, dwarf and pygmy sperm whales, and river dolphins). Most schooling species have characteristic school sizes (although they can vary somewhat area to area). For example, rough-toothed dolphins typically occur in groups of 10–20, pilot whales occur in schools of dozens, and some oceanic dolphins (Stenella spp., Delphinus spp.) regularly occur in groups of hundreds or thousands.

School size correlates with feeding habits: species that form large schools are almost all shallow-diving species that feed mainly on schooling prey, whereas those which occur in school sizes of 25 or fewer tend to be (a) deep-diving species and feed mainly on larger squids or (b) coastal species feeding on dispersed prey. School size also correlates with predation pressure; large cetaceans, presumably subject to lower predation pressure than small species, occur only in small groups, whereas small cetaceans, subject to higher predation pressure, occur in schools whose size correlates with openness of habitat: the more open, the larger the school size. School size should correlate with resource availability and will affect reproductive strategies, although the nature of these relationships remains largely unexplored.

Although most schools are monospecific, several species regularly occur in mixed-species schools. Some of these associations appear to be opportunistic: bottlenose dolphins, e.g., have been recorded to occur with over 20 different species of whales and dolphins. Other associations appear to be more prescribed: spotted and spinner dolphins regularly occur together in mixed schools. Risso's, Pacific white-sided (Lagenorhynchus obliquidens), and northern right whale (Lissodelphis borealis) dolphins are commonly found in association. The nature of these interactions (e.g., why these species-specific associations occur, how these species avoid competition) is unknown.

Communities and Coexistence

Studies of communities typically focus on identifying member species and their interactions and then address mechanisms for their coexistence. These kinds of studies comprise a large part of the ecological knowledge for many terrestrial species. In contrast, very little is known about this aspect of cetacean ecology.

There are regularly occurring species assemblages. For example, pantropical spotted and spinner dolphins are frequently found in mixed-species schools in association with yellow fin tuna (Thunnus albacares) and are accompanied by large and speciose flocks of seabirds; this association is particularly prevalent in the eastern tropical Pacific, as opposed to other tropical oceans. There are variations in typical co-occurrence patterns. In the Gulf of Mexico, e.g., five species of Stenella coexist in a relatively small area, more Stenella species than any other tropical ocean. The nature of the interactions between species in these assemblages, why they associate, and the reasons for variations in community membership patterns are almost completely unknown.

Coexisting species, particularly those that are closely related or have similar ecological roles, potentially compete for resources. An often cited example is the southern ocean, where the relative abundances of cetaceans, pinnipeds, and seabirds, all krill consumers, have been reported to have changed between pre- and post-whaling years. One plausible explanation is competitive release: the decrease in biomass of cetacean predators released a huge prey base of krill to pinnipeds and seabirds, both of which were able to increase in abundance.

Ecological theory states that stable communities of coexisting species must differ in resource utilization in some way: through prey species or size specialization, differential habitat use, or diet pattern. Such niche partitioning is fairly clear for cetaceans on a broad scale. For example, there are species that feed on fish and those that feed on squid. There are species feeding in shallow water and those that feed at depth. Some cetaceans feed at night and others during the day.

On a smaller scale, one of the best known examples of niche partitioning is for baleen whales. In this group, there is a fair degree of prey specialization that presumably allows for niche partitioning in areas of sympatry. Blue whales feed almost entirely on euphausiids; fi n whales and humpbacks feed mainly on fishes but take euphausiids when they are abundant; and right whales and sei whales feed mainly on copepods. Odontocetes provide additional possible examples. Bottlenose, short-beaked common (Delphinus delphis), pantropical spotted dolphins, and harbor porpoise exhibit diet specialization among age, sex, and reproductive class, although this diet specialization could be due to differing energy requirements. Aside from these examples, very little is known about how, or if, cetaceans partition resources.

Ultimately, to understand community structure, the mechanisms by which species partition resources, not merely the presence of differences in resource use, are of principal interest. The question then becomes, given that there are differences, what mechanisms can explain them? Community ecologists have identified interference and exploitative competition, mutualism, morphological or physiological factors, and habitat structure as potential mechanisms for maintaining resource utilization differences. This is an area that remains almost completely unexplored for cetaceans and the communities in which they are found.

Role of Cetaceans in Marine Ecosystems

Australasian gannets (Sula serrator) feed on small-schooling fish
(e.g., pilchard, Sardinops neopilchardus) which have been herded to the
surface by dusky dolphins (Lagenorhynchus obscurus) in Admiralty Bay, New Zealand.

What role do cetaceans play in marine ecosystems and what is their significance? Most cetaceans are apex predators. As such, they take tons of prey from the ecosystem. In so doing, it seems likely that cetaceans affect the life history strategies and population biology of their prey, as well as organisms at other trophic levels that interact in various ways with these prey. Little is known about the details of these dynamics, although this may be the most significant way in which cetaceans impact marine ecosystems.

More specific effects have been documented. For example, benthic feeders such as gray whales alter habitat by regularly turning over substrate (between 9% and 27% of the benthos in the northern Bering Sea) and therefore, significantly affect the species composition of benthic communities. Feeding cetaceans provide feeding opportunities for seabirds by driving prey to the surface, sometimes injuring or disorienting it. In one study up to 87% of all feeding individuals from four seabird species in the Bering Sea associated with gray whale mud plumes. Large whales dying at sea may sink to the bottom and provide rare but superabundant food and habitat for deep-water species. There is evidence that mollusc communities may have specialized on these resources for the past 35 million years, and some speculate that whale carcasses may have been instrumental in the dispersal of hydrothermal-vent faunas. Feces of some cetaceans, particularly large whales in areas of low productivity, may play a significant role in nutrient cycling. Cetaceans are host to a variety of commensal or parasitic species; in some cases (Cyamid whale lice), these species are completely dependent on cetaceans through all life stages.

Macroecology

Macroecology is a branch of ecology that attempts to characterize the relationships between organisms and their environment by increasing the spatial and temporal scale of investigation. Macroecologists ask questions about the relationships between abundance, distribution, and diversity of individuals, populations, or species and incorporate principles of biogeography, paleobiology, systematics, and the earth sciences in their search for pattern. Macroecology is a relatively recent field but has resulted in significant and novel insights. For example, the concept of species richness hotspots, the underlying mechanisms that produce them and the ecological and conservation implications are all products of the macroecological perspective. These advances pertain largely to terrestrial systems but there are notable exceptions. Worldwide patterns of tuna and billfish diversity indicate that there are clear hotspots, areas of high species richness, that appear to hold in general for other taxa and trophic levels, and are functions of oceanography. The significance of these patterns and their relevance to cetaceans is unknown, but a promising field of investigation.

BALEEN WHALE

Baleen whales are widely distributed and diverse parvorder of carnivorous marine mammals. Mysticeti comprise the families Balaenidae (right and bowhead whales), Balaenopteridae (rorquals), Cetotheriidae (the pygmy right whale), and Eschrichtiidae (the gray whale). There are currently 15 species of baleen whales. While cetaceans were historically thought to have descended from mesonychids, (which would place them outside the order Artiodactyla), molecular evidence supports them as a clade of even-toed ungulates (Artiodactyla). Baleen whales split from toothed whales (Odontoceti) around 34 million years ago.

Baleen whales range in size from the 20 ft (6 m) and 6,600 lb (3,000 kg) pygmy right whale to the 102 ft (31 m) and 190 t (210 short tons) blue whale, the largest known animal to have ever existed. They are sexually dimorphic. Baleen whales can have streamlined or large bodies, depending on the feeding behavior, and two limbs that are modified into flippers. Though not as flexible and agile as seals, baleen whales can swim very fast, with the fastest able to travel at 23 miles per hour (37 km/h). Baleen whales use their baleen plates to filter out food from the water by either lunge-feeding or skim-feeding. Baleen whales have fused neck vertebrae, and are unable to turn their head at all. Baleen whales have two blowholes. Some species are well adapted for diving to great depths. They have a layer of fat, or blubber, under the skin to keep warm in the cold water.

Although baleen whales are widespread, most species prefer the colder waters of the Arctic and Antarctic. Gray whales are specialized for feeding on bottom-dwelling crustaceans. Rorquals are specialized at lunge-feeding, and have a streamlined body to reduce drag while accelerating. Right whales skim-feed, meaning they use their enlarged head to effectively take in a large amount of water and sieve the slow-moving prey. Males typically mate with more than one female (polygyny), although the degree of polygyny varies with the species. Male strategies for reproductive success vary between performing ritual displays (whale song) or lek mating. Calves are typically born in the winter and spring months and females bear all the responsibility for raising them. Mothers fast for a relatively long period of time over the period of migration, which varies between species. Baleen whales produce a number of vocalizations, notably the songs of the humpback whale.

The meat, blubber, baleen, and oil of baleen whales have traditionally been used by the indigenous peoples of the Arctic. Once relentlessly hunted by commercial industries for these products, cetaceans are now protected by international law. However, the North Atlantic right whale is ranked endangered by the International Union for Conservation of Nature. Besides hunting, baleen whales also face threats from marine pollution and ocean acidification. It has been speculated that man-made sonar results in strandings. They have rarely been kept in captivity, and this has only been attempted with juveniles or members of one of the smallest species.

Baleen whales are cetaceans classified under the parvorder Mysticeti, and consist of four extant families: Balaenidae (right whales), Balaenopteridae (rorquals), Cetotheriidae (pygmy right whale), and Eschrichtiidae (gray whale). Balaenids are distinguished by their enlarged head and thick blubber, while rorquals and gray whales generally have a flat head, long throat pleats, and are more streamlined than Balaenids. Rorquals also tend to be longer than the latter. Cetaceans (whales, dolphins, and porpoises) and artiodactyls are now classified under the order Cetartiodactyla, often still referred to as Artiodactyla (given that the cetaceans are deeply nested with the artiodactyls). The closest living relatives to baleen whales are toothed whales both from the infraorder Cetacea.

Balaenidae consists of two genera: Eubalaena (right whales) and Balaena (the bowhead whale, B. mysticetus). Balaenidae was thought to have consisted of only one genus until studies done through the early 2000s reported that bowhead whales and right whales are morphologically (different skull shape) and phylogenically different. According to a study done by H. C. Rosenbaum (of the American Museum of Natural History) and colleagues, the North Pacific (E. japonica) and Southern right (E. australis) whales are more closely related to each other than to the North Atlantic right whale (E. glacialis).

Rorquals consist of two genera (Balaenoptera and Megaptera) and nine species: the fin whale (B. physalus), the Sei whale (B. borealis), Bryde's whale (B. brydei), Eden's whale (B. edeni), the blue whale (B. musculus), the common minke whale (B. acutorostrata), the Antarctic minke whale (B. bonaerensis), Omura's whale (B. omurai), and the humpback whale (M. novaeangliae). In a 2012 review of cetacean taxonomy, Alexandre Hassanin (of the Muséum National d'Histoire Naturelle) and colleagues suggested that, based on phylogenic criteria, there are four extant genera of rorquals. They recommend that the genus Balaenoptera be limited to the fin whale, have minke whales fall under the genus Pterobalaena, and have Rorqualus contain the Sei whale, Bryde's whale, Eden's whale, the blue whale, and Omura's whale.

Cetotheriidae consists of only one living member: the pygmy right whale (Caperea marginata). The first descriptions date back to the 1840s of bones and baleen plates resembling a smaller version of the right whale, and was named Balaena marginata. In 1864, it was moved into the genus Caperea after a skull of another specimen was discovered. Six years later, the pygmy right whale was classified under the family Neobalaenidae. Despite its name, the pygmy right whale is more genetically similar to rorquals and gray whales than to right whales. A study published in 2012, based on bone structure, moved the pygmy right whale from the family Neobalaenidae to the family Cetotheriidae, making it a living fossil; Neobalaenidae was elevated down to subfamily level as Neobalaeninae.

Eschrichtiidae consists of only one living member: the gray whale (Eschrichtius robustus). The two populations, one in the Sea of Okhotsk and Sea of Japan and the other in the Mediterranean Sea and East Atlantic, are thought to be genetically and physiologically dissimilar. However, DNA analysis by studies, such as the one by Takeshi Sasaki and colleagues, indicates certain rorquals, such as the humpback whale, Megaptera novaeangliae, and the fin whale, Balaenoptera physalus, are more closely related to the gray whale than they are to some other rorquals, such as the minke whale, Balaenoptera acutorostrata.

Differences between Families

Baleen whales vary considerably in size and shape, depending on their feeding behavior.

Rorquals use throat pleats to expand their mouths, which allow them to feed more effectively. However, rorquals need to build up water pressure in order to expand their mouths, leading to a lunge-feeding behavior. Lunge-feeding is where a whale rams a bait ball (a swarm of small fish) at high speed. Rorquals generally have streamlined physiques to reduce drag in the water while doing this. Balaenids rely on their huge heads, as opposed to the rorquals' throat pleats, to feed effectively. This feeding behavior allows them to grow very big and bulky, without the necessity for a streamlined body. They have callosities, unlike other whales, with the exception of the bowhead whale. Rorquals have a higher proportion of muscle tissue and tend to be negatively buoyant, whereas right whales have a higher proportion of blubber and are positively buoyant. Gray whales are easily distinguished from other extant cetaceans by their sleet-gray color, dorsal ridges (knuckles on the back), and their gray-white scars left from parasites. As with the rorquals, their throat pleats increase the capacity of their throats, allowing them to filter larger volumes of water at once. Gray whales are bottom-feeders, meaning they sift through sand to get their food. They usually turn on their sides, scoop up sediment into their mouths and filter out benthic creatures like amphipods, which leave noticeable marks on their heads. The pygmy right whale is easily confused with minke whales because of their similar characteristics, such as their small size, dark gray tops, light gray bottoms, and light eye-patches.

Eschrichtiidae.

Eubalaena, Balaenidae.

Balaena, Balaenidae.

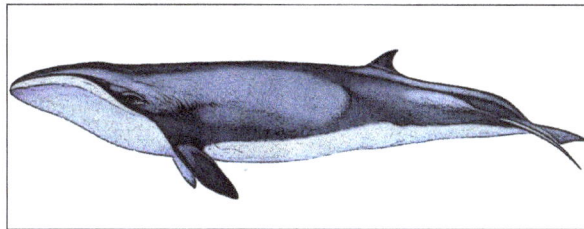
Cetotheriidae.

Evolutionary History

Mysticeti split from Odontoceti (toothed whales) 26 to 17 million years ago during the Eocene. Their evolutionary link to archaic toothed cetaceans (Archaeoceti) remained unknown until the extinct Janjucetus hunderi was discovered in the early 1990s in Victoria, Australia. Like a modern baleen whale, Janjucetus had baleen in its jaw and had very little biosonar capabilities. However, its jaw also contained teeth, with incisors and canines built for stabbing and molars and premolars built for tearing. These early mysticetes were exceedingly small compared to modern baleen whales, with species like Mammalodon measuring no greater than 10 feet (3 m). It is thought that their size increased with their dependence on baleen. However, the discovery of a skull of the toothed Llanocetus, the second-oldest mysticete, yielded a total length of 8 meters (26 ft), indicating filter feeding was not a driving feature in mysticete evolution. The discovery of Janjucetus and others like it suggests that baleen evolution went through several transitional phases. Species like Mammalodon colliveri had little to no baleen, while later species like Aetiocetus weltoni had both baleen and teeth, suggesting they had limited filter feeding capabilities; later genera like Cetotherium had no teeth in their mouth, meaning they were fully dependent on baleen and could only filter feed. However, the 2018 discovery of the toothless Maiabalaena indicates some lineages evolved toothlessness before baleen.

Restoration of Janjucetus hunderi.

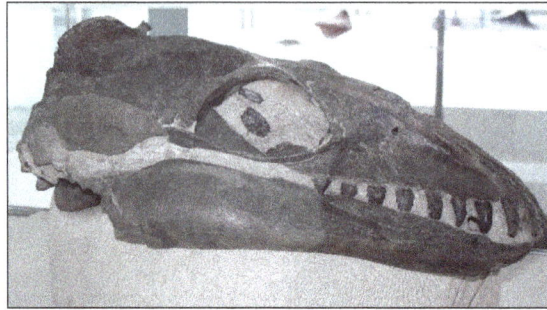
Archaeomysticetes, like Janjucetus, had teeth.

Mystacodon selenensis is the earliest mysticete, dating back to 37 to 33 million years ago (mya) in the Late Eocene, and, like other early toothed mysticetes, or "archaeomysticetes", M. selenensis had heterodont dentition used for suction feeding. Archaeomysticetes from the Oligocene are the Mammalodontidae (Mammalodon and Janjucetus) from Australia. They were small with short-ened rostra, and a primitive dental formula (3.1.4.33.1.4.3). In baleen whales, it is thought that enlarged mouths adapted for suction feeding evolved before specializations for bulk filter feeding. In the toothed Oligocene mammalodontid Janjucetus, the symphysis is short and the mouth en-larged, the rostrum is wide, and the edges of the maxillae are thin, indicating an adaptation for suction feeding. The aetiocetid Chonecetus still had teeth, but the presence of a groove on the inte-rior side of each mandible indicates the symphysis was elastic, which would have enabled rotation of each mandible, an initial adaptation for bulk feeding like in modern mysticetes.

The first toothless ancestors of Mysticetes appeared before the first radiation in the late Oligocene. Eomysticetus and others like it showed no evidence in the skull of echolocation abilities, suggesting they mainly relied on their eyesight for navigation. The eomysticetes had long, flat rostra that lacked teeth and had blowholes located halfway up the dorsal side of the snout. Though the palate is not well-preserved in these specimens, they are thought to have had baleen and been filter feeders. Mio-cene baleen whales were preyed upon by larger predators like killer sperm whales and Megalodon.

Miocene mysticetes were often hunted by
megalodon and killer sperm whales.

The lineages of rorquals and right whales split almost 20 mya. It is unknown where this occurred, but it is generally believed that they, like their descendants, followed plankton migrations. These primitive mysticetes had lost their heterodont dentition in favor of baleen, and are believed to have lived on a specialized benthic, plankton, or copepod diet like modern mysticetes. Mysticetes experienced their first radiation in the mid-Miocene. It is thought this radiation was caused by global climate change and major tectonic activity when Antarctica and Australia separated from each other, creating the Antarctic Circumpolar Current. Balaenopterids grew bigger during this time, with species like Balaenoptera sibbaldina perhaps rivaling the blue whale in terms of size, though many other studies disagree that any mysticetes grew that large in the miocene.

The increase in size is likely due to shifts in climate that have resulted in seasonally shifting accumulations of plankton in various parts of the world, necessitating more efficiency traveling over long distances between widely distributed prey sources which also resulted in a lower metabolic rate, and feeding on baitballs. A 2017 analysis of body size based on data from the fossil record and modern mysticetes indicates that the evolution of gigantism in baleen whales is a recent phenomenon, estimated to have occurred within the past 3.1 Ma. Before 4.5 million years ago, few mysticetes exceeded 10 meters (33 feet) in length; the two largest miocene species were less than 13 m in length. The initial evolution of baleen and filter feeding long preceded the evolution of gigantic body size, indicating the evolution of novel feeding mechanisms did not cause the evolution of gigantism. The creation of the Antarctic circumpolar current and its effects on global climate patterns is excluded as being causal for the same reason. Gigantism also was preceded by divergence of different mysticete lineages, meaning multiple lineages arrived at large size independently. It is possible the Plio-Pleistocene increase in seasonally intense upwellings causing high-prey-density zones led to gigantism.

Anatomy

Motion

A humpback whale skeleton. Notice how the jaw is split into two.

When swimming, baleen whales rely on their flippers for locomotion in a wing-like manner similar to penguins and sea turtles. Flipper movement is continuous. While doing this, baleen whales use their tail fluke to propel themselves forward through vertical motion while using their flippers for steering, much like an otter. Some species leap out of the water, which may allow them to travel faster. Because of their great size, right whales are not flexible or agile like dolphins, and none can move their neck because of the fused cervical vertebrae; this sacrifices speed for stability in the water. The hind legs are enclosed inside the body, and are thought to be vestigial organs. However, a 2014 study suggests that the pelvic bone serves as support for whale genitalia.

Rorquals, needing to build speed to feed, have several adaptions for reducing drag, including a streamlined body; a small dorsal fin, relative to its size; and lack of external ears or long hair. The fin whale, the fastest among baleen whales, can travel at 23 miles per hour (37 km/h). While feeding, the rorqual jaw expands to a volume that can be bigger than the whale itself; to do this, the mouth inflates. The inflation of the mouth causes the cavum ventrale, the throat pleats on the underside stretching to the navel, to expand, increasing the amount of water that the mouth can store. The mandible is connected to the skull by dense fibers and cartilage (fibrocartilage), allowing the jaw to swing open at almost a 90° angle. The mandibular symphysis is also fibrocartilaginous, allowing the jaw to bend which lets in more water. To prevent stretching the mouth too far, rorquals have a sensory organ located in the middle of the jaw to regulate these functions.

External Anatomy

Paired blowholes of a humpback and the V-shaped blow of a right whale.

Baleen whales have two flippers on the front, near the head. Like all mammals, baleen whales breathe air and must surface periodically to do so. Their nostrils, or blowholes, are situated at the top of the cranium. Baleen whales have two blowholes, as opposed to toothed whales which have one. These paired blowholes are longitudinal slits that converge anteriorly and widen posteriorly, which causes a V-shaped blow. They are surrounded by a fleshy ridge that keeps water away while the whale breathes. The septum that separates the blowholes has two plugs attached to it, making the blowholes water-tight while the whale dives.

Like other mammals, the skin of baleen whales has an epidermis, a dermis, a hypodermis, and connective tissue. The epidermis, the pigmented layer, is 0.2 inches (5 mm) thick, along with connective tissue. The epidermis itself is only 0.04 inches (1 mm) thick. The dermis, the layer underneath the epidermis, is also thin. The hypodermis, containing blubber, is the thickest part of the skin and functions as a means to conserve heat. Right whales have the thickest hypodermis of any cetacean, averaging 20 inches (51 cm), though, as in all whales, it is thinner around openings (such as the blowhole) and limbs. Blubber may also be used to store energy during times of fasting. The connective tissue between the hypodermis and muscles allows only limited movement to occur between them. Unlike in toothed whales, baleen whales have small hairs on the top of their head, stretching from the tip of the rostrum to the blowhole, and, in right whales, on the chin. Like other marine mammals, they lack sebaceous and sweat glands.

Baleen

The baleen of baleen whales are keratinous plates. They are made of a calcified, hard α-keratin material, a fiber-reinforced structure made of intermediate filaments (proteins). The degree of calcification varies between species, with the sei whale having 14.5% hydroxyapatite, a mineral that coats teeth and bones, whereas minke whales have 1–4% hydroxyapatite. In most mammals, keratin structures, such as wool, air-dry, but aquatic whales rely on calcium salts to form on the plates to stiffen them. Baleen plates are attached to the upper jaw and are absent in the mid-jaw, forming two separate combs of baleen. The plates decrease in size as they go further back into the jaw; the largest ones are called the "main baleen plates" and the smallest ones are called the "accessory plates". Accessory plates taper off into small hairs.

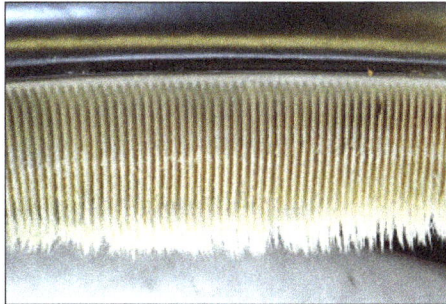

Accessory baleen plates taper off into small hairs.

Sexual Dimorphism

Unlike other whales (and most other mammals), the females are larger than the males. Sexual dimorphism is usually reversed, with the males being larger, but the females of all baleen whales are usually five percent larger than males. Sexual dimorphism is also displayed through whale song, notably in humpback whales where the males of the species sing elaborate songs. Male right whales have bigger callosities than female right whales. The males are generally more scarred than females which is thought to be because of aggression during mating season.

Internal Systems

The unique lungs of baleen whales are built to collapse under the pressure instead of resisting the pressure which would damage the lungs, enabling some, like the fin whale, to dive to a depth of −1,540 feet (−470 m). The whale lungs are very efficient at extracting oxygen from the air, usually 80%, whereas humans only extract 20% of oxygen from inhaled air. Lung volume is relatively low compared to terrestrial mammals because of the inability of the respiratory tract to hold gas while diving. Doing so may cause serious complications such as embolism. Unlike other mammals, the lungs of baleen whales lack lobes and are more sacculated. Like in humans, the left lung is smaller than the right to make room for the heart. To conserve oxygen, blood is rerouted from pressure-tolerant-tissue to internal organs, and they have a high concentration of myoglobin which allows them to hold their breath longer.

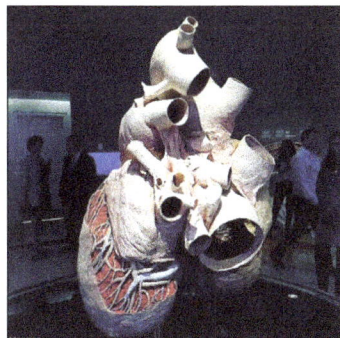

The heart of a blue whale with a person standing next to it.

The heart of baleen whales functions similarly to other mammals, with the major difference being the size. The heart can reach 1,000 pounds (454 kg), but is still proportional to the whale's size. The muscular wall of the ventricle, which is responsible for pumping blood out of the heart, can be 3 to 5 inches (7.6 to 12.7 cm) thick. The aorta, an artery, can be 0.75 inches (1.9 cm) thick. Their resting heart rate is 60 to 140 beats per minute (bpm), as opposed to the 60 to 100 bpm in humans. When diving, their heart rate will drop to 4 to 15 bpm to conserve oxygen. Like toothed whales, they have a dense network of blood vessels (rete mirabile) which prevents heat-loss. Like in most mammals, heat is lost in their extremities, so, in baleen whales, warm blood in the arteries is surrounded by veins to prevent heat loss during transport. As well as this, heat inevitably given off by the arteries warms blood in the surrounding veins as it travels back into the core. This is otherwise known as countercurrent exchange. To counteract overheating while in warmer waters, baleen whales reroute blood to the skin to accelerate heat-loss. They have the largest blood corpuscles (red and white blood cells) of any mammal, measuring $4.1 \times 10-4$ inches (10 μm) in diameter, as opposed to human's $2.8 \times 10-4$-inch (7.1 μm) blood corpuscles.

When sieved from the water, food is swallowed and travels through the esophagus where it enters a three-chambered-stomach. The first compartment is known as the fore-stomach; this is where food gets ground up into an acidic liquid, which is then squirted into the main stomach. Like in humans, the food is mixed with hydrochloric acid and protein-digesting enzymes. Then, the partly digested food is moved into the third stomach, where it meets fat-digesting enzymes, and is then mixed with an alkaline liquid to neutralize the acid from the fore-stomach to prevent damage to the intestinal tract. Their intestinal tract is highly adapted to absorb the most nutrients from food; the walls are folded and contain copious blood vessels, allowing for a greater surface area over which digested food and water can be absorbed. Baleen whales get the water they need from their food; however, the salt content of most of their prey (invertebrates) are similar to that of seawater, whereas the salt content of a whale's blood is considerably lower (three times lower) than that of seawater. The whale kidney is adapted to excreting excess salt; however, while producing urine more concentrated than seawater, it wastes a lot of water which must be replaced.

Baleen whales have a relatively small brain compared to their body mass. Like other mammals, their brain has a large, folded cerebrum, the part of the brain responsible for memory and processing sensory information. Their cerebrum only makes up about 68% of their brain's weight, as opposed to human's 83%. The cerebellum, the part of the brain responsible for balance and coordination, makes up 18% of their brain's weight, compared to 10% in humans, which is probably due to the great degree of control necessary for constantly swimming. Necropsies on the brains of gray whales revealed iron oxide particles, which may allow them to find magnetic north like a compass.

Unlike most animals, whales are conscious breathers. All mammals sleep, but whales cannot afford to become unconscious for long because they may drown. They are believed to exhibit unihemispheric slow-wave sleep, in which they sleep with half of the brain while the other half remains active. This behavior was only documented in toothed whales until footage of a humpback whale sleeping (vertically) was shot in 2014.

It is largely unknown how baleen whales produce sound because of the lack of a melon and vocal cords. In a 2007 study, it was discovered that the larynx had U-shaped folds which are thought to be similar to vocal cords. They are positioned parallel to air flow, as opposed to the perpendicular vocal cords of terrestrial mammals. These may control air flow and cause vibrations. The walls of the larynx are able to contract which may generate sound with support from the arytenoid carti-lages. The muscles surrounding the larynx may expel air rapidly or maintain a constant volume while diving.

Senses

Their eyes are relatively small for their size.

The eyes of baleen whales are relatively small for their size and are positioned near the end of the mouth. This is probably because they feed on slow or immobile prey, combined with the fact that most sunlight does not pass 30 feet (9.1 m), and hence they do not need acute vision. A whale's eye is adapted for seeing both in the euphotic and aphotic zones by increasing or decreasing the pupil's size to prevent damage to the eye. As opposed to land mammals which have a flattened lens, whales have a spherical lens. The retina is surrounded by a reflective layer of cells (tapetum lucidum), which bounces light back at the retina, enhancing eyesight in dark areas. However, light is bent more near the surface of the eye when in air as opposed to water; consequently, they can see much better in the air than in the water. The eyeballs are protected by a thick outer layer to prevent abrasions, and an oily fluid (instead of tears) on the surface of the eye. Baleen whales appear to have limited color vision, as they lack S-cones.

The mysticete ear is adapted for hearing underwater, where it can hear sound frequencies as low as 7 Hz and as high as 22 kHz. It is largely unknown how sound is received by baleen whales. Unlike in toothed whales, sound does not pass through the lower jaw. The auditory meatus is blocked by connective tissue and an ear plug, which connects to the eardrum. The inner-ear bones are con-tained in the tympanic bulla, a bony capsule. However, this is attached to the skull, suggesting that vibrations passing through the bone is important. Sinuses may reflect vibrations towards the co-chlea. It is known that when the fluid inside the cochlea is disturbed by vibrations, it triggers sen-sory hairs which send electrical current to the brain, where vibrations are processed into sound.

Baleen whales have a small, yet functional, vomeronasal organ. This allows baleen whales to detect chemicals and pheromones released by their prey. It is thought that 'tasting' the water is important for finding prey and tracking down other whales. They are believed to have an impaired sense of smell due to the lack of the olfactory bulb, but they do have an olfactory tract. Baleen whales have few if any taste buds, suggesting they have lost their sense of taste. They do retain salt-receptor taste-buds suggesting that they can taste saltiness.

Behavior

Migration

Most species of baleen whale migrate long distances from high latitude waters during spring and summer months to more tropical waters during winter months. This migration cycle is repeated annually. The gray whale has the longest recorded migration of any mammal, with one traveling 14,000 miles (23,000 km) from the Sea of Okhotsk to the Baja Peninsula.

It is thought that plankton blooms dictate where whales migrate. Many baleen whales feed on the massive plankton blooms that occur in the cold, nutrient rich waters of polar regions during the sunny spring and summer months. Baleen whales generally then migrate to calving grounds in tropical waters during the winter months when plankton populations are low. Migration is hypothesized to benefit calves in a number of ways. Newborns, born with underdeveloped blubber, would likely otherwise be killed by the cold polar temperatures. Migration to warmer waters may also reduce the risk of calves being predated on by killer whales.

Migratory movements may also reflect seasonally shifting patterns of productivity. California blue whales are hypothesized to migrate between dense patches of prey, moving from central California in the summer and fall, to the Gulf of California in the winter, to the central Baja California Pacific coast in spring.

Foraging

Humpback whales lunge-feeding in the
course of bubble net fishing.

All modern mysticetes are obligate filter feeders, using their baleen to strain small prey items (including small fish, krill, copepods, and zooplankton) from seawater. Despite their carnivorous diet, a 2015 study revealed they house gut flora similar to that of terrestrial herbivores. Different kinds of prey are found in different abundances depending on location, and each type of whale is adapted to a specialized way of foraging.

There are two types of feeding behaviors: skim-feeding and lunge-feeding, but some species do both depending on the type and amount of food. Lunge-feeders feed primarily on euphausiids (krill), though some smaller lunge feeders (e.g. minke whales) also prey on schools of fish. Skim-feeders, like bowhead whales, feed upon primarily smaller plankton such as copepods. They feed alone or in small groups. Baleen whales get the water they need from their food, and their kidneys excrete excess salt.

The lunge-feeders are the rorquals. To feed, lunge-feeders expand the volume of their jaw to a volume bigger than the original volume of the whale itself; to do this, the mouth inflates to expand the mouth. The inflation of the mouth causes the throat pleats to expand, increasing the amount of water that the mouth can store. Just before they ram the baitball, the jaw swings open at almost a 90° angle and bends which lets in more water. To prevent stretching the mouth too far, rorquals have a sensory organ located in the middle of the jaw to regulate these functions. Then they must decelerate. This process takes a lot of mechanical work, and is only energy-effective when used against a large baitball. Lunge feeding is more energy intensive than skim-feeding due to the acceleration and deceleration required.

The skim-feeders are right whales, gray whales, pygmy right whales, and sei whales (which also lunge feed). To feed, skim-feeders swim with an open mouth, filling it with water and prey. Prey must occur in sufficient numbers to trigger the whale's interest, be within a certain size range so that the baleen plates can filter it, and be slow enough so that it cannot escape. The "skimming" may take place on the surface, underwater, or even at the ocean's bottom, indicated by mud occasionally observed on right whales' bodies. Gray whales feed primarily on the ocean's bottom, feeding on benthic creatures.

Foraging efficiency for both lunge feeding and continuous ram filter feeding is highly dependent upon prey density. The efficiency of a blue whale lunge is approximately 30 times higher at krill densities of 4.5 kg/m3 than at low krill densities of 0.15 kg/m3. Baleen whale have been observed seeking out highly specific areas within the local environment in order to forage at the highest density prey aggregations.

Predation and Parasitism

Orange whale lice on a right whale.

Baleen whales, primarily juveniles and calves, are preyed on by killer whales. It is thought that annual whale migration occurs to protect the calves from the killer whales. There have also been reports of a pod of killer whales attacking and killing an adult bowhead whale, by holding down its flippers, covering the blowhole, and ramming and biting until death. Generally, a mother and calf pair, when faced with the threat of a killer whale pod, will either fight or flee. Fleeing only occurs in species that can swim away quickly, the rorquals. Slower whales must fight the pod alone or with a small family group. There has been one report of a shark attacking and killing a whale calf. This occurred in 2014 during the sardine run when a shiver of dusky sharks attacked a humpback whale calf. Usually, the only shark that will attack a whale is the cookie cutter shark, which leaves a small, non-fatal bite mark.

Many parasites latch onto whales, notably whale lice and whale barnacles. Almost all species of whale lice are specialized towards a certain species of whale, and there can be more than one species per whale. Whale lice eat dead skin, resulting in minor wounds in the skin. Whale louse infestations are especially evident in right whales, where colonies propagate on their callosities. Though not a parasite, whale barnacles latch onto the skin of a whale during their larval stage. However, in doing so it does not harm nor benefit the whale, so their relationship is often labeled as an example of commensalism. Some baleen whales will deliberately rub themselves on substrate to dislodge parasites. Some species of barnacle, such as Conchoderma auritum and whale barnacles, attach to the baleen plates, though this seldom occurs. A species of copepod, Balaenophilus unisetus, inhabits baleen plates of whales in tropical waters. A species of Antarctic diatom, Cocconeis ceticola, forms a film on the skin, which takes a month to develop; this film causes minor damage to the skin. They are also plagued by internal parasites such as stomach worms, cestodes, nematodes, liver flukes, and acanthocephalans.

Reproduction and Development

Female right whale with calf.

Before reaching adulthood, baleen whales grow at an extraordinary rate. In the blue whale, the largest species, the fetus grows by some 220 lb (100 kg) per day just before delivery, and by 180 lb (80 kg) per day during suckling. Before weaning, the calf increases its body weight by 17 t (17 long tons; 19 short tons) and grows from 23 to 26 ft (7 to 8 m) at birth to 43 to 52 ft (13 to 16 m) long. When it reaches sexual maturity after 5–10 years, it will be 66 to 79 ft (20 to 24 m) long and possibly live as long as 80–90 years. Calves are born precocial, needing to be able to swim to the surface at the moment of their birth.

Most rorquals mate in warm waters in winter to give birth almost a year later. A 7-to-11 month lactation period is normally followed by a year of rest before mating starts again. Adults normally start reproducing when 5–10 years old and reach their full length after 20–30 years. In the smallest rorqual, the minke whale, 10 ft (3 m) calves are born after a 10-month pregnancy and weaning lasts until it has reached about 16 to 18 ft (5 to 5.5 m) after 6–7 months. Unusual for a baleen whale, female minkes (and humpbacks) can become pregnant immediately after giving birth; in most species, there is a two-to-three-year calving period. In right whales, the calving interval is usually three years. They grow very rapidly during their first year, after which they hardly increase in size for several years. They reach sexual maturity when 43 to 46 ft (13 to 14 m) long. Baleen whales are K-strategists, meaning they raise one calf at a time, have a long life-expectancy, and a low infant mortality rate. Some 19th century harpoons found in harvested bowheads indicate this species can live more than 100 years. Baleen whales are promiscuous, with none showing pair bonds. They are polygynous, in that a male may mate with more than one female. The scars on male whales suggest they fight for the right to mate with females during breeding season, somewhat similar to lek mating.

Baleen whales have fibroelastic (connective tissue) penises, similar to those of artiodactyls. The tip of the penis, which tapers toward the end, is called the pars intrapraeputialis or terminal cone. The blue whale has the largest penis of any organism on the planet, typically measuring 8–10 feet (2.4–3.0 m). Accurate measurements of the blue whale are difficult to take because the whale's erect length can only be observed during mating. The penis on a right whale can be up to 2.7 m (8.9 ft) – the testes, at up to 2 m (6.6 ft) in length, 78 cm (2.56 ft) in diameter, and weighing up to 525 lb (238 kg), are also the largest of any animal on Earth.

Whale Song

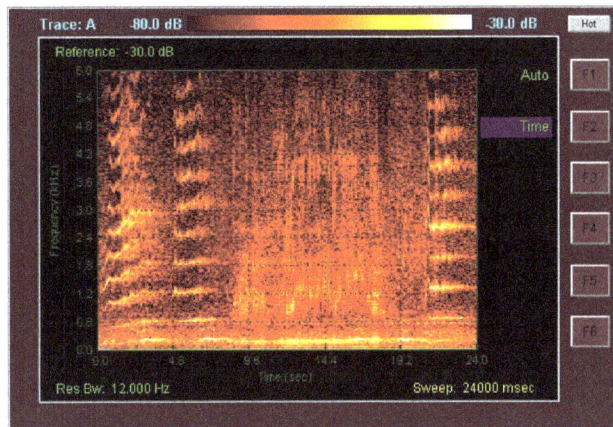

Spectrogram of humpback whale vocalizations: detail is shown for the first 24 seconds of the 37-second recording "Singing Humpbacks". The whale songs are heard before and after a set of echolocation clicks in the middle.

All baleen whales use sound for communication and are known to "sing", especially during the breeding season. Blue whales produce the loudest sustained sounds of any animals: their low-frequency (about 20 Hz) moans can last for half a minute, reach almost 190 decibels, and be heard hundreds of kilometers away. Adult male humpbacks produce the longest and most complex songs; sequences of moans, groans, roars, sighs, and chirps sometimes lasting more than ten minutes are repeated for hours. Typically, all humpback males in a population sing the same song over a breeding season, but the songs change slightly between seasons, and males in one population have been observed adapting the song from males of a neighboring population over a few breeding seasons.

Intelligence

Unlike their toothed whale counterparts, baleen whales are hard to study because of their immense size. Intelligence tests such as the mirror test cannot be done because their bulk and lack of body language makes a reaction impossible to be definitive. However, studies on the brains of humpback whales revealed spindle cells, which, in humans, control theory of mind. Because of this, it is thought that baleen whales, or at least humpback whales, have consciousness.

PAKICETUS

Pakicetus is an extinct genus of amphibious cetacean of the family Pakicetidae, which was endemic to Pakistan during the Eocene. The vast majority of paleontologists regard it as the most basal whale.

Based on the skull sizes of specimens, and to a lesser extent on composite skeletons, species of Pakicetus are thought to have been 1 metre (3 ft 3 in) to 2 metres (6 ft 7 in) in length.

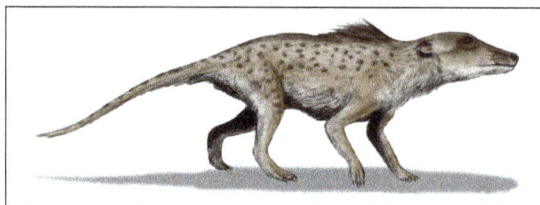

P. inachus life restoration.

Pakicetus looked very different from modern cetaceans, and its body shape more resembled those of land dwelling, hoofed mammals. Unlike all later cetaceans, it had four fully functional, long legs. Pakicetus had a long snout; a typical complement of teeth that included incisors, canines, premolars, and molars; a distinct and flexible neck; and a very long and robust tail. As in most land mammals, the nose was situated at the tip of the snout.

Reconstructions of pakicetids that followed the discovery of composite skeletons often depicted them with fur; however, given their relatively close relationships with hippos, they may have had sparse body hair.

The first fossil found consisted of an incomplete skull with a skull cap and a broken mandible with some teeth. Based on the detail of the teeth, the molars suggest that the animal could rend and tear flesh. Wear, in the form of scrapes on the molars indicated that Pakicetus ground its teeth as it chewed its food. Because of the toothwear, Pakicetus is thought to have eaten fish and small animals. The teeth also suggest that Pakicetus had herbivorous and omnivorous ancestors.

Palaeobiology

Possible Semi-aquatic Nature

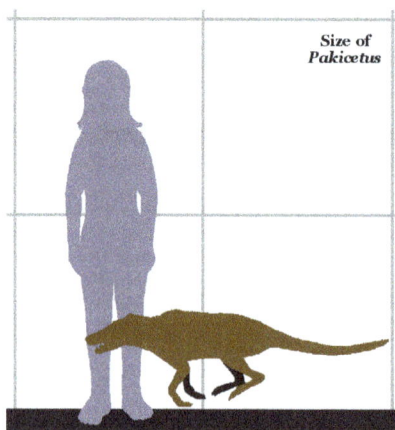

Size of
Pakicetus

Size of Pakicetus, compared to a human.

It was illustrated on the cover of Science as a semiaquatic, vaguely crocodile-like mammal, diving after fish. Somewhat more complete skeletal remains were discovered in 2001, prompting the view that Pakicetus was primarily a land animal about the size of a wolf, and very similar in form to the related mesonychids. Thewissen et al. 2001 wrote that "Pakicetids were terrestrial mammals, no more amphibious than a tapir."

However, Thewissen et al. 2009 argued that "the orbits of these cetaceans were located close together on top of the skull, as is common in aquatic animals that live in water but look at emerged objects. Just like Indohyus, limb bones of pakicetids are osteosclerotic, also suggestive of aquatic habitat" (since heavy bones provide ballast). "This peculiarity could indicate that Pakicetus could stand in water, almost totally immersed, without losing visual contact with the air."

Sensory Capabilities

The Pakicetus skeleton reveals several details regarding the creature's unique senses, and provides a newfound ancestral link between terrestrial and aquatic animals. As previously mentioned, the Pakicetus' upward-facing eye placement was a significant indication of its habitat. Even more so, however, was its auditory abilities. Like all other cetaceans, Pakicetus had a thickened skull bone known as the auditory bulla, which was specialized for underwater hearing. Cetaceans also all categorically exhibit a large mandibular foramen within the lower jaw, which holds a fat pack and extends towards the ear, both of which are also associated with underwater hearing. "Pakicetus is the only cetacean in which the mandibular foramen is small, as is the case in all terrestrial animals. It thus lacked the fat pad, and sounds reached its eardrum following the external auditory meatus as in terrestrial mammals. Thus the hearing mechanism of Pakicetus is the only known intermediate between that of land mammals and aquatic cetaceans." With both the auditory and visual senses in mind, as well as the typical diet of Pakicetus, one might assume that the creature was able to attack both aquatic and terrestrial prey from a low vantage point.

Classification

Pakicetus was originally described as being a mesonychid, but later research reclassified it as an early cetacean due to characteristic features of the inner ear found only in cetaceans, (namely, the large auditory bulla is formed from the ectotympanic bone only), it was recognized as the earliest member of the family Pakicetidae. It was originally believed to be descended from mesonychids, according to Gingerich & Russell 1981. However, the redescription of the primitive, semi-aquatic artiodactyl Indohyus, and the discovery of its cetacean-like inner ear, simultaneously put an end to the idea that whales were descended from mesonychids, while demonstrating that Pakicetus, and all other cetaceans, are artiodactyls. Thus, Pakicetus represents a transitional taxon between extinct land mammals and modern cetaceans.

References

- Janik, Vincent (2014). "Cetacean vocal learning and communication". Current Opinion in Neurobiology. 28: 60–65. Doi:10.1016/j.conb.2014.06.010. PMID 25057816

- Cetacean, cetacean: britannica.com, Retrieved 8 January, 2019

- Holmes, Bob. "A life spent chasing down how whales evolved". New Scientist. New Scientist. Retrieved December 2, 2018

- Derr, Mark (May 2001). "Mirror test". New York Times. Retrieved 3 August 2015

- E. Fish, Frank (2002). "Balancing Requirements for Stability and Maneuverability in Cetaceans". Integrative and Comparative Biology. 42 (1): 85–93. Doi:10.1093/icb/42.1.85. PMID 21708697

- Fox, David (2001). "Balaenoptera physalus (fin whale)". Animal Diversity Web. University of Michigan Museum of Zoology. Retrieved 22 October 2006

- Cooper, Lisa Noelle; Thewissen, J.G.M; Hussain, S. T (2009). "New middle Eocene archaeocetes (Cetacea: Mammalia) from the Kuldana Formation of northern Pakistan". Journal of Vertebrate Paleontology. 29 (4): 1289–1299. Doi:10.1671/039.029.0423

4

Pinnipeds

Pinnipeds are a diverse category of carnivorous, fin-footed, semiaquatic marine mammals. They have streamlined bodies and four limbs which are evolved into flippers. Some of the different types of pinnipeds are sea lions, eared seals and walruses. The chapter closely examines the key features of these types of pinnipeds to provide an extensive understanding of the subject.

Pinnipeds are fin-footed mammals comprising seals, sea lions, and the walrus. Pinnipeds live only in rich marine environments and a few inland or tropical freshwater systems.

Shaped like torpedoes, pinnipeds have wide torsos and narrower hindquarters. They are extremely awkward on land but swift and graceful in the water. Their slitlike nostrils can be closed underwater, and externally the ears are either small or entirely absent. All have short fur, the walrus having almost none, and the tail is vestigial. Length ranges from 1.1 to 6.5 metres (3.6 to 21 feet), and weights range from about 30 kg (66 pounds) in some female fur seals to 3,700 kg in male elephant seals (genus Mirounga).

Northern elephant seal (Mirounga angustirostris).

Fur seals and sea lions (family Otariidae) and walrus (family Odobenidae) use their sizable forelimbs for propulsion, whereas true, or earless, seals (family Phocidae) use mostly the hind flippers. Despite the great size of some species, all are agile and easily capture fish in open water. Pinnipeds are visual predators, and, even though they may lack external ears, they generally have excellent hearing, especially underwater. All pinnipeds also have sensitive whiskers that help them detect prey. Diet is strictly carnivorous, but pinnipeds eat a variety of prey ranging in size from krill, which is filtered from the water by the complex cheek teeth, to, in the case of the leopard seal (Hydrurga leptonyx), penguins and even other pinnipeds. Most, however, rely primarily on fish, crustaceans (crabs, lobsters, shrimp), cephalopods (squid and octopus), and mollusks (shellfish).

Pinnipeds such as the Ross seal (Ommatophoca rossii) may be solitary at certain times of the year, but most are usually gregarious, much more so than terrestrial carnivores. During the breeding season more than a million may congregate on an island. Males are sometimes larger than females, and among elephant seals males can be five times larger. Their greater size allows them to better defend harems of many females. In other pinnipeds the sexes are similar in size. Most males mate with multiple females, but some pinnipeds are monogamous. Mating and birth occur on coastal land or ice or on ice floes. Implantation of the fertilized eggs is delayed, with the result that gestation can last from 8 to 15 months. Young seals are called pups, and single offspring are the rule, with twins occurring rarely. At birth pups are often a different colour than their parents. After the breeding season most pinnipeds are pelagic (open-sea dwellers), traveling long distances either alone or in small groups, though some species do not migrate from the breeding grounds. The young mature in less than 6 years (longer in walrus), and some species can live more than 30 years in the wild. They are preyed upon by sharks, killer whales, leopard seals, and polar bears. Pinnipeds are also hunted by humans for their skin, meat, and fat (blubber). Walrus are also hunted for their ivory tusks.

The German naturalist Johann Karl Wilhelm Illiger was the first to recognize the pinnipeds as a distinct taxonomic unit; in 1811 he gave the name Pinnipedia to both a family and an order. American zoologist Joel Asaph Allen reviewed the world's pinnipeds in an 1880 monograph, History of North American pinnipeds, a monograph of the walruses, sea-lions, sea-bears and seals of North America. He traced the history of names, gave keys to families and genera, described North American species and provided synopses of species in other parts of the world. In 1989, Annalisa Berta and colleagues proposed the unranked clade Pinnipedimorpha to contain the fossil genus Enaliarctos and modern seals as a sister group. Pinnipeds belong to the order Carnivora and the suborder Caniformia (known as dog-like carnivorans). Pinnipedia was historically considered its own suborder under Carnivora. Of the three extant families, the Otariidae and Odobenidae are grouped in the superfamily Otarioidea, while the Phocidae belong to the superfamily Phocoidea.

Otariids are also known as eared seals due to the presence of pinnae. These animals rely on their well-developed fore-flippers to propel themselves through the water. They can also turn their hind-flippers forward and "walk" on land. The anterior end of an otariid's frontal bones extends between the nasal bones, and the supraorbital foramen is large and flat horizontally. The supraspinatous fossas are divided by a "secondary spine" and the bronchi are divided anteriorly. Otariids consist of two types: sea lions and fur seals. Sea lions are distinguished by their rounder snouts and shorter, rougher pelage, while fur seals have more pointed snouts, longer fore-flippers and thicker fur coats that include an undercoat and guard hairs. The former also tend to be larger than the latter. Five genera and seven species (one now extinct) of sea lion are known to exist, while two genera and nine species of fur seal exist. While sea lions and fur seals have historically been considered separate subfamilies (Otariinae and Arctocephalinae respectively), a 2001 genetic study found that the northern fur seal is more closely related to several sea lion species. This is supported by a 2006 molecular study that also found that the Australian sea lion and New Zealand sea lion are more closely related to Arctocephalus than to other sea lions.

Odobenidae consists of only one living member: the modern walrus. This animal is easily distinguished from other extant pinnipeds by its larger size (exceeded only by the elephant seals), nearly hairless skin and long upper canines, known as tusks. Like otariids, walruses are capable of turning their hind-flippers forward and can walk on land. When moving in water, the walrus relies on its hind-flippers for locomotion, while its fore-flippers are used for steering. In addition, the walrus lacks external ear flaps. Walruses have pterygoid bones that are broad and thick, frontal bones that are V-shaped at the anterior end and calcaneuses with pronounced tuberosity in the middle.

Phocids are known as true or "earless" seals. These animals lack external ear flaps and are incapable of turning their hind-flippers forward, which makes them more cumbersome on land. In water, true seals swim by moving their hind-flippers and lower body from side to side. Phocids have thickened mastoids, enlarged entotympanic bones, everted pelvic bones and massive ankle bones. They also lack supraorbital processes on the frontal and have underdeveloped calcaneal tubers. A 2006 molecular study supports the division of phocids into two monophyletic subfamilies: Monachinae, which consists of Mirounga, Monachini and Lobodontini; and Phocinae, which includes Pusa, Phoca, Halichoerus, Histriophoca, Pagophilus, Erignathus and Cystophora.

In a 2012 review of pinniped taxonomy, Berta and Morgan Churchill suggested that, based on morphological and genetic criteria, there are 33 extant species and 29 subspecies of pinnipeds, although five of the latter lack sufficient support to be conclusively considered subspecies. They recommend that the genus Arctocephalus be limited to Arctocephalus pusillus, and they resurrected the name Arctophoca for several species and subspecies formerly placed in Arctocephalus. More than 50 fossil species have been described.

Restoration of Puijila darwini.

One popular hypothesis suggested that pinnipeds are diphyletic (descended from two ancestral lines), with walruses and otariids sharing a recent common ancestor with bears and phocids sharing one with Musteloidea. However, morphological and molecular evidence support a monophyletic origin. Nevertheless, there is some dispute as to whether pinnipeds are more closely related to bears or musteloids, as some studies support the former theory and others the latter. Pinnipeds split from other caniforms 50 million years ago (mya) during the Eocene. Their evolutionary link to terrestrial mammals was unknown until the 2007 discovery of Puijila darwini in early Miocene deposits in Nunavut, Canada. Like a modern otter, Puijila had a long tail, short limbs and webbed feet instead of flippers. However, its limbs and shoulders were more robust and Puijila likely had been a quadrupedal swimmer—retaining a form of aquatic locomotion that give rise to the major swimming types employed by modern pinnipeds. The researchers who found Puijila placed it in a clade with Potamotherium (traditionally considered a mustelid) and Enaliarctos. Of the three, Puijila was the least specialized for aquatic life. The discovery of Puijila in a lake deposit suggests that pinniped evolution went through a freshwater transitional phase.

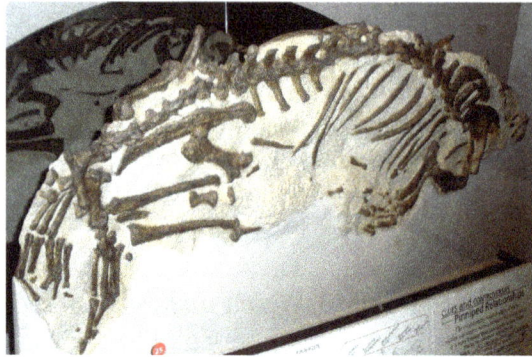
Fossil of Enaliarctos.

Enaliarctos, a fossil species of late Oligocene/early Miocene (24–22 Mya) California, closely resembled modern pinnipeds; it was adapted to an aquatic life with a flexible spine, and limbs modified into flippers. Its teeth were adapted for shearing (like terrestrial carnivorans), and it may have stayed near shore more often than its extant relatives. Enaliarctos was capable of swimming with both the fore-flippers and hind-flippers, but it may have been more specialized as a fore-flipper swimmer. One species, Enaliarctos emlongi, exhibited notable sexual dimorphism, suggesting that this physical characteristic may have been an important driver of pinniped evolution. A closer relative of extant pinnipeds was Pteronarctos, which lived in Oregon 19–15 mya. As in modern seals, Pteroarctos had an orbital wall that was not limited by certain facial bones (like the jugal or lacrimal bone), but was mostly shaped by the maxilla.

The ancestors of the Otarioidea and Phocoidea diverged 33 mya. The Phocidae are likely to have descended from the extinct family Desmatophocidae in the North Atlantic. Desmatophocids lived 23–10 Mya and had elongated skulls, fairly large eyes, cheekbones connected by a mortised structure and rounded cheek teeth. They also were sexually dimorphic and may have been capable of propelling themselves with both the foreflippers and hindflippers.

Fossil skull cast of Desmatophoca oregonensis
from the extinct Desmatophocidae.

Phocids are known to have existed for at least 15 million years,and molecular evidence supports a divergence of the Monachinae and Phocinae lineages 22 Mya. The fossil monachine Monotherium and phocine Leptophoca were found in southeastern North America. The deep split between the lineages of Erignathus and Cystophora 17 Mya suggests that the phocines migrated eastward and northward from the North Atlantic. The genera Phoca and Pusa could have arisen when a phocine lineage traveled from the Paratethys Sea to the Arctic Basin and subsequently went eastward. The ancestor of the Baikal seal migrated into Lake Baikal from the Arctic (via the Siberian ice sheet) and became isolated there. The Caspian seal's ancestor became isolated as the Paratethys shrank, leaving the animal in a small remnant sea, the Caspian Sea. The monochines diversified southward. Monachus emerged in the Mediterranean and migrated to the Caribbean and then the central North Pacific. The two extant elephant seal species diverged close to 4 mya after the Panamanian isthmus was formed. The lobodontine lineage emerged around 9 mya and colonized the southern ocean in response to glaciation.

The lineages of Otariidae and Odobenidae split almost 28 Mya. Otariids originated in the North Pacific. The earliest fossil Pithanotaria, found in California, is dated to 11 mya. The Callorhinus lineage split earlier at 16 mya. Zalophus, Eumetopias and Otaria diverged next, with the latter colonizing the coast of South America. Most of the other otariids diversified in the Southern Hemisphere. The earliest fossils of Odobenidae—Prototaria of Japan and Proneotherium of Oregon—date to 18–16 Mya. These primitive walruses had much shorter canines and lived on a fish diet rather than a specialized mollusk diet like the modern walrus. Odobenids further diversified in the middle and late Miocene. Several species had enlarged upper and lower canines. The genera Valenictus and Odobenus developed elongated tusks. The lineage of the modern walrus may have spread from the North Pacific to the Caribbean (via the Central American Seaway) 8–5 Mya and subsequently made it to the North Atlantic and returned to the North Pacific via the Arctic 1 mya.

Alternatively, this lineage may have spread from the North Pacific to the Arctic and subsequently the North Atlantic during the Pleistocene.

Anatomy and Physiology

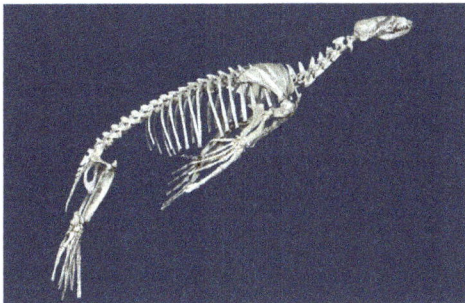

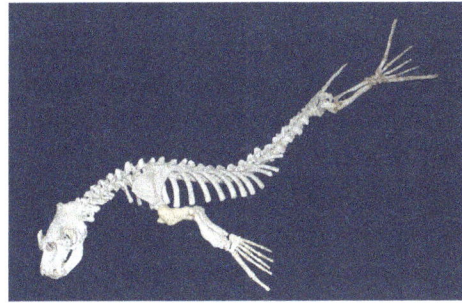

Skeleton of California sea lion (top) and southern elephant seal.

Pinnipeds have streamlined, spindle-shaped bodies with reduced or non-existent external ear flaps, rounded heads, flexible necks, limbs modified into flippers, and small tails. Pinniped skulls have large eye orbits, short snouts and a constricted interorbital region. They are unique among carnivorans in that their orbital walls are significantly shaped by the maxilla and are not limited by certain facial bones. Compared to other carnivorans, their teeth tend to be fewer in number (especially incisors and back molars), are pointed and cone-shaped, and lack carnassials. The walrus has unique upper canines that are elongated into tusks. The mammary glands and genitals of pinnipeds can retract into the body.

Pinnipeds range in size from the 1 m (3 ft 3 in) and 45 kg (99 lb) Baikal seal to the 5 m (16 ft) and 3,200 kg (7,100 lb) southern elephant seal. Overall, they tend to be larger than other carnivorans; the southern elephant seal is the largest carnivoran. Several species have male-biased sexual dimorphism that correlates with the degree of polygyny in a species: highly polygynous species like elephant seals are extremely sexually dimorphic, while less polygynous species have males and females that are closer in size. In lobodontine seals, females are slightly larger than males. Males of sexually dimorphic species also tend to have secondary sex characteristics, such as the prominent proboscis of elephant seals, the inflatable red nasal membrane of hooded seals and the thick necks and manes of otariids. Despite a correlation between size dimorphism and the degree of polygyny, some evidence suggests that size differences between the sexes originated due to ecological differences and prior to the development of polygyny.

Male and female South American sea lions,
showing sexual dimorphism.

Almost all pinnipeds have fur coats, the exception being the walrus, which is only sparsely covered. Even some fully furred species (particularly sea lions) are less haired than most land mammals. In species that live on ice, young pups have thicker coats than adults. The individual hairs on the coat, known collectively as lanugo, can trap heat from sunlight and keep the pup warm. Pinnipeds are typically countershaded, and are darker colored dorsally and lighter colored ventrally, which serves to eliminate shadows caused by light shining over the ocean water. The pure white fur of harp seal pups conceals them in their Arctic environment. Some species, such as ribbon seals, ringed seals and leopard seals, have patterns of contrasting light and dark coloration. All fully furred species molt; phocids molt once a year, while otariids gradually molt all year. Seals have a layer of subcutaneous fat known as blubber that is particularly thick in phocids and walruses. Blubber serves both to keep the animals warm and to provide energy and nourishment when they are fasting. It can constitute as much as 50% of a pinniped's body weight. Pups are born with only a thin layer of blubber, but some species compensate for this with thick lanugos.

Pinnipeds have a simple stomach that is similar in structure to terrestrial carnivores. Most species have neither a cecum nor a clear demarcation between the small and large intestines; the large intestine is comparatively short and only slightly wider than the latter. Small intestine lengths range from 8 times (California sea lion) to 25 times (elephant seal) the body length. The length of the intestine may be an adaptation to frequent deep diving, as the increased volume of the digestive tract serves as an extended storage compartment for partially digested food during submersion. Pinnipeds do not have an appendix. As in most marine mammals, the kidneys are divided into small lobes and can effectively absorb water and filter out excess salt.

Locomotion

Harbor seal left and California sea lion swimming. The former swims with its hind-flippers, the latter with its fore- flippers.

Pinnipeds have two pairs of flippers on the front and back, the fore-flippers and hind-flippers. The elbows and ankles are enclosed within the body. Pinnipeds tend to be slower swimmers than cetaceans, typically cruising at 5–15 km (9–28 km/h; 6–17 mph) compared to around 20 kn (37 km/h; 23 mph) for several species of dolphin. Seals are more agile and flexible, and some otariids, such as the California sea lion, are capable of bending their necks backwards far enough to reach their hind-flippers, allowing them to make dorsal turns. Pinnipeds have several adaptions for reducing drag. In addition to their streamlined bodies, they have smooth networks of muscle bundles in their skin that may increase laminar flow and make it easier for them to slip through water. They also lack arrector pili, so their fur can be streamlined as they swim.

When swimming, otariids rely on their fore-flippers for locomotion in a wing-like manner similar to penguins and sea turtles. Fore-flipper movement is not continuous, and the animal glides between each stroke. Compared to terrestrial carnivorans, the fore-limbs of otariids are reduced in length, which gives the locomotor muscles at the shoulder and elbow joints greater mechanical advantage; the hind-flippers serve as stabilizers. Phocids and walruses swim by moving their hind-flippers and lower body from side to side, while their fore-flippers are mainly used for steering. Some species leap out of the water, which may allow then to travel faster. In addition, sea lions are known to "ride" waves which probably helps them decrease their energy usage.

Pinnipeds can move around on land, though not as well as terrestrial animals. Otariids and walruses are capable of turning their hind-flippers forward and under the body so they can "walk" on all fours. The fore-flippers move in a transverse, rather than a sagittal fashion. Otariids rely on the movements of their heads and necks more than their hind-flippers during terrestrial locomotion. By swinging their heads and necks, otariids create momentum while they are moving. Sea lions have been recorded climbing up flights of stairs. Phocids are less agile on land. They cannot pull their hind-flippers forward, and move on land by lunging, bouncing and wiggling while their fore-flippers keep them balanced. Some species use their fore-flippers to pull themselves forward. Terrestrial locomotion is easier for phocids on ice, as they can sled along.

Senses

The eyes of pinnipeds are relatively large for their size and are positioned near the front of the head. One exception is the walrus, whose smaller eyes are located on the sides of its head. This is because it feeds on immobile bottom dwelling mollusks and hence does not need acute vision. A seal's eye is adapted for seeing both underwater and in air. The lens is mostly spherical, and much of the retina is equidistant from the lens center. The cornea has a flattened center where refraction is nearly equal in both water and air. Pinnipeds also have very muscular and vascularized irises. The well-developed dilator muscle gives the animals a great range in pupil dilation. When contracted, the pupil is typically pear-shaped, although the bearded seal's is more diagonal. In species that live in shallow water, such as harbor seals and California sea lions, dilation varies little, while the deep-diving elephant seals have much greater variation.

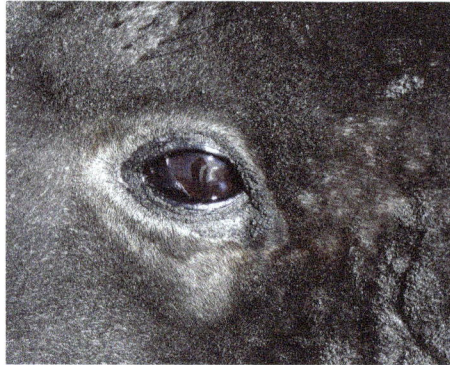

Light reflection on an elephant seal eye.

On land, pinnipeds are near-sighted in dim light. This is reduced in bright light, as the retracted pupil reduces the lens and cornea's ability to bend light. They also have a well-developed tapetum lucidum, a reflecting layer that increases sensitivity by reflecting light back through the rods. This helps them see in low-light conditions. Ice-living seals like the harp seal have corneas that can tolerate high levels of ultraviolet radiation typical of bright, snowy environments. As such, they do not suffer snow blindness. Pinnipeds appear to have limited color vision, as they lack S-cones. Flexible eye movement has been documented in seals. The extraocular muscles of the walrus are well developed. This and its lack of orbital roof allow it to protrude its eyes and see in both frontal and dorsal directions. Seals release large amounts of mucus to protect their eyes. The corneal epithelium is keratinized and the sclera is thick enough to withstand the pressures of diving. As in many mammals and birds, pinnipeds possess nictitating membranes.

Frontal view of brown fur seal head.

The pinniped ear is adapted for hearing underwater, where it can hear sound frequencies at up to 70,000 Hz. In air, hearing is somewhat reduced in pinnipeds compared to many terrestrial mammals. While they are capable of hearing a wide range of frequencies (e.g. 500 to 32,000 Hz in the northern fur seal, compared to 20 to 20,000 Hz in humans), their airborne hearing sensitivity is weaker overall. One study of three species—the harbor seal, California sea lion and northern elephant seal—found that the sea lion was best adapted for airborne hearing, the harbor seal was equally capable of hearing in air and water, and the elephant seal was better adapted for underwater hearing. Although pinnipeds have a fairly good sense of smell on land, it is useless underwater as their nostrils are closed.

Vibrissae of walrus.

Pinnipeds have well-developed tactile senses. Their mystacial vibrissae have ten times the innervation of terrestrial mammals, allowing them to effectively detect vibrations in the water. These vibrations are generated, for example, when a fish swims through water. Detecting vibrations is useful when the animals are foraging and may add to or even replace vision, particularly in darkness. Harbor seals have been observed following varying paths of another seal that swam ahead several minutes before, similar to a dog following a scent trail, and even to discriminate the species and the size of the fish responsible for the trail. Blind ringed seals have even been observed successfully hunting on their own in Lake Saimaa, likely relying on their vibrissae to gain sensory information and catch prey.

Unlike terrestrial mammals, such as rodents, pinnipeds do not move their vibrissae over an object when examining it but instead extend their moveable whiskers and keep them in the same position. By holding their vibrissae steady, pinnipeds are able to maximize their detection ability. The vibrissae of phocids are undulated and wavy while otariid and walrus vibrissae are smooth. Research is ongoing to determine the function, if any, of these shapes on detection ability. The vibrissa's angle relative to the flow, not the shape, however, seems to be the most important factor. The vibrissae of some otariids grow quite long—those of the Antarctic fur seal can reach 41 cm (16 in). Walruses have the most vibrissae, at 600–700 individual hairs. These are important for detecting their prey on the muddy sea floor. In addition to foraging, vibrissae may also play a role in navigation; spotted seals appear to use them to detect breathing holes in the ice.

Diving Adaptations

Diving Weddell seals.

Before diving, pinnipeds typically exhale to empty their lungs of half the air and then close their nostrils and throat cartilages to protect the trachea. Their unique lungs have airways that are highly reinforced with cartilaginous rings and smooth muscle, and alveoli that completely deflate during deeper dives. While terrestrial mammals are generally unable to empty their lungs, pinnipeds can reinflate their lungs even after complete respiratory collapse. The middle ear contains sinuses that probably fill with blood during dives, preventing middle ear squeeze. The heart of a seal is moderately flattened to allow the lungs to deflate. The trachea is flexible enough to collapse under pressure. During deep dives, any remaining air in their bodies is stored in the bronchioles and trachea, which prevents them from experiencing decompression sickness, oxygen toxicity and nitrogen narcosis. In addition, seals can tolerate large amounts of lactic acid, which reduces skeletal muscle fatigue during intense physical activity.

The main adaptations of the pinniped circulatory system for diving are the enlargement and increased complexity of veins to increase their capacity. Retia mirabilia form blocks of tissue on the inner wall of the thoracic cavity and the body periphery. These tissue masses, which contain extensive contorted spirals of arteries and thin-walled veins, act as blood reservoirs that increase oxygen stores for use during diving. As with other diving mammals, pinnipeds have high amounts of hemoglobin and myoglobin stored in their blood and muscles. This allows them to stay submerged for long periods of time while still having enough oxygen. Deep-diving species such as elephant seals have blood volumes that make up to 20% of their body weight. When diving, they reduce their heart rate and maintain blood flow only to the heart, brain and lungs. To keep their blood pressure stable, phocids have an elastic aorta that dissipates some energy of each heartbeat.

Thermoregulation

Northern elephant seal resting in water.

Pinnipeds conserve heat with their large and compact body size, insulating blubber and fur, and high metabolism. In addition, the blood vessels in their flippers are adapted for countercurrent exchange. Veins containing cool blood from the body extremities surround arteries, which contain warm blood received from the core of the body. Heat from the arterial blood is transferred to the blood vessels, which then recirculate blood back to the core. The same adaptations that conserve heat while in water tend to inhibit heat loss when out of water. To counteract overheating, many species cool off by flipping sand onto their backs, adding a layer of cool, damp sand that enhances heat loss. The northern fur seal pants to help stay cool, while monk seals often dig holes in the sand to expose cooler layers to rest in.

Sleep

Pinnipeds spend many months at a time at sea, so they must sleep in the water. Scientists have recorded them sleeping for minutes at a time while slowly drifting downward in a belly-up orientation. Like other marine mammals, seals sleep in water with half of their brain awake so that they can detect and escape from predators. When they are asleep on land, both sides of their brain go into sleep mode.

Distribution and Habitat

Walrus on ice off Alaska. This species has a discontinuous distribution around the Arctic Circle.

Living pinnipeds mainly inhabit polar and subpolar regions, particularly the North Atlantic, the North Pacific and the Southern Ocean. They are entirely absent from Indo-Malayan waters. Monk seals and some otariids live in tropical and subtropical waters. Seals usually require cool, nutrient-rich waters with temperatures lower than 20 °C (68 °F). Even those that live in warm or tropical climates live in areas that become cold and nutrient rich due to current patterns. Only monk seals live in waters that are not typically cool or rich in nutrients. The Caspian seal and Baikal seal are found in large landlocked bodies of water (the Caspian Sea and Lake Baikal respectively).

As a whole, pinnipeds can be found in a variety of aquatic habitats, including coastal water, open ocean, brackish water and even freshwater lakes and rivers. Most species inhabit coastal areas, though some travel offshore and feed in deep waters off oceanic islands. The Baikal seal is the only freshwater species, though some ringed seals live in freshwater lakes in Russia close to the Baltic sea. In addition, harbor seals may visit estuaries, lakes and rivers and sometimes stay as long as a year. Other species known to enter freshwater include California sea lions and South American sea lions. Pinnipeds also use a number of terrestrial habitats and substrates, both continental and island. In temperate and tropical areas, they haul out on to sandy and pebble beaches, rocky shores, shoals, mud flats, tide pools and in sea caves. Some species also rest on man-made structures, like piers, jetties, buoys and oil platforms. Pinnipeds may move further inland and rest in sand dunes or vegetation, and may even climb cliffs. Polar-living species haul out on to both fast ice and drift ice.

Behavior

Pinnipeds have an amphibious lifestyle; they spend most of their lives in the water, but haul out to mate, raise young, molt, rest, thermoregulate or escape from aquatic predators. Several species are known to migrate vast distances, particularly in response to extreme environmental changes, like El Niño or changes in ice cover. Elephant seals stay at sea 8–10 months a year and migrate between breeding and molting sites. The northern elephant seal has one of the longest recorded migration distances for a mammal, at 18,000–21,000 km (11,000–13,000 mi). Phocids tend to migrate more than otariids. Traveling seals may use various features of their environment to reach their destination including geomagnetic fields, water and wind currents, the position of the sun and moon and the taste and temperature of the water.

Harbor seal hauled out on rock.

Pinnipeds may dive during foraging or to avoid predators. When foraging, Weddell seals typically dive for less than 15 minutes to depths of around 400 m (1,300 ft) but can dive for as long as 73 minutes and to depths of up to 600 m (2,000 ft). Northern elephant seals commonly dive 350–650 m (1,150–2,130 ft) for as long as 20 minutes. They can also dive 1,259–4,100 m (4,131–13,451 ft) and for as long as 62 minutes. The dives of otariids tend to be shorter and less deep. They typically last 5–7 minutes with average depths to 30–45 m (98–148 ft). However, the New Zealand sea lion has been recorded diving to a maximum of 460 m (1,510 ft) and a duration of 12 minutes. Walruses do not often dive very deep, as they feed in shallow water.

Pinnipeds have lifespans averaging 25–30 years. Females usually live longer, as males tend to fight and often die before reaching maturity. The longest recorded lifespans include 43 years for a wild female ringed seal and 46 years for a wild female grey seal. The age at which a pinniped sexually matures can vary from 2–12 years depending on the species. Females typically mature earlier than males.

Foraging and Predation

Steller sea lion with white sturgeon.

All pinnipeds are carnivorous and predatory. As a whole, they mostly feed on fish and cephalopods, followed by crustaceans and bivalves, and then zooplankton and endothermic ("warm-blooded") prey like sea birds. While most species are generalist and opportunistic feeders, a few are specialists. Examples include the crabeater seal, which primarily eats krill, the ringed seal, which eats mainly crustaceans, the Ross seal and southern elephant seal, which specialize on squid, and the bearded seal and walrus, which feed on clams and other bottom-dwelling invertebrates. Pinnipeds may hunt solitarily or cooperatively. The former behavior is typical when hunting non-schooling fish, slow-moving or immobile invertebrates or endothermic prey. Solitary foraging species usually exploit coastal waters, bays and rivers. An exception to this is the northern elephant seal, which feeds on fish at great depths in the open ocean. In addition, walruses feed solitarily but are often near other walruses in small or large groups that may surface and dive in unison. When large schools of fish or squid are available, pinnipeds such as certain otariids hunt cooperatively in large groups, locating and herding their prey. Some species, such as California and South American sea lions, may forage with cetaceans and sea birds.

Seals typically consume their prey underwater where it is swallowed whole. Prey that is too large or awkward is taken to the surface to be torn apart. The leopard seal, a prolific predator of penguins, is known to violently swing its prey back and forth until it is dead. The elaborately cusped teeth of filter-feeding species, such as crabeater seals, allow them to remove water before they swallow their planktonic food. The walrus is unique in that it consumes its prey by suction feeding, using its tongue to suck the meat of a bivalve out of the shell. While pinnipeds mostly hunt in the water, South American sea lions are known to chase down penguins on land. Some species may swallow stones or pebbles for reasons not understood. Though they can drink seawater, pinnipeds get most of their fluid intake from the food they eat.

Leopard seal capturing emperor penguin.

Pinnipeds themselves are subject to predation. Most species are preyed on by the killer whale or orca. To subdue and kill seals, orcas continuously ram them with their heads, slap them with their tails and fling them in the air. They are typically hunted by groups of 10 or fewer whales, but they are occasionally hunted by larger groups or by lone individuals. Pups are more commonly taken by orcas, but adults can be targeted as well. Large sharks are another major predator of pinnipeds—usually the great white shark but also the tiger shark and mako shark. Sharks usually attack by ambushing them from below. The prey usually escapes, and seals are often seen with shark-inflicted wounds. Otariids typically have injuries in the hindquarters, while phocids usually have injuries on the forequarters. Pinnipeds are also targeted by terrestrial and pagophilic predators. The polar bear is well adapted for hunting Arctic seals and walruses, particularly pups. Bears are known to use sit-and-wait tactics as well as active stalking and pursuit of prey on ice or water. Other terrestrial predators include cougars, brown hyenas and various species of canids, which mostly target the young.

Killer whale hunting a Weddell seal.

Pinnipeds lessen the chance of predation by gathering in groups. Some species are capable of inflicting damaging wounds on their attackers with their sharp canines—an adult walrus is capable of killing polar bears. When out at sea, northern elephant seals dive out of the reach of surface-hunting orcas and white sharks. In the Antarctic, which lacks terrestrial predators, pinniped species spend more time on the ice than their Arctic counterparts. Arctic seals use more breathing holes per individual, appear more restless when hauled out, and rarely defecate on the ice. Ringed seals build dens underneath fast ice for protection.

Interspecific predation among pinnipeds does occur. The leopard seal is known to prey on numerous other species, especially the crabeater seal. Leopard seals typically target crabeater pups, which form an important part of their diet from November to January. Older crabeater seals commonly bear scars from failed leopard seal attacks; a 1977 study found that 75% of a sample of 85 individual crabeaters had these scars. Walruses, despite being specialized for feeding on bottom-dwelling invertebrates, occasionally prey on Arctic seals. They kill their prey with their long tusks and eat their blubber and skin. Steller sea lions have been recorded eating the pups of harbor seals, northern fur seals and California sea lions. New Zealand sea lions feed on pups of some fur seal species, and the South American sea lion may prey on South American fur seals.

Reproductive Behavior

Walrus herd on ice floe.

The mating system of pinnipeds varies from extreme polygyny to serial monogamy. Of the 33 species, 20 breed on land, and the remaining 13 breed on ice. Species that breed on land are usually polygynous, as females gather in large aggregations and males are able to mate with them as well as defend them from rivals. Polygynous species include elephant seals, grey seals and most otariids. Land-breeding pinnipeds tend to mate on islands where there are fewer terrestrial predators.

Few islands are favorable for breeding, and those that are tend to be crowded. Since the land they breed on is fixed, females return to the same sites for many years. The males arrive earlier in the season and wait for them. The males stay on land and try to mate with as many females as they can; some of them will even fast. If a male leaves the beach to feed, he will likely lose mating opportunities and his dominance. Polygynous species also tend to be extremely sexual dimorphic in favor of males. This dimorphism manifests itself in larger chests and necks, longer canines and denser fur—all traits that help males in fights for females. Increased body weight in males increases the length of time they can fast due to the ample energy reserves stored in the blubber. Larger males also likely enjoy access to feeding grounds that smaller ones are unable to access due to their lower thermoregulatory ability and decreased energy stores. In some instances, only the largest males are able to reach the furthest deepest foraging grounds where they enjoy maximum energetic yields that are unavailable to smaller males and females.

Other seals, like the walrus and most phocids, breed on ice with copulation usually taking place in the water (a few land-breeding species also mate in water). Females of these species tend to aggregate less. In addition, since ice is less stable than solid land, breeding sites change location each year, and males are unable to predict where females will stay during the breeding season. Hence polygyny tends to be weaker in ice-breeding species. An exception to this is the walrus, where females form dense aggregations perhaps due to their patchy food sources. Pinnipeds that breed on fast ice tend to cluster together more than those that breed on drift ice. Some of these species are serially monogamous, including the harp seal, crabeater seal and hooded seal. Seals that breed on ice tend to have little or no sexual dimorphism. In lobodontine seals, females are slightly longer than males. Walruses and hooded seals are unique among ice-breeding species in that they have pronounced sexual dimorphism in favor of males.

Northern fur seal breeding colony.

Adult male pinnipeds have several strategies to ensure reproductive success. Otariids establish territories containing resources that attract females, such as shade, tide pools or access to water. Territorial boundaries are usually marked by natural breaks in the substrate, and some may be fully or partially underwater. Males defend their territorial boundaries with threatening vocalizations and postures, but physical fights are usually avoided. Individuals also return to the same territorial site each breeding season. In certain species, like the Steller sea lion and northern fur seal, a dominant male can maintain a territory for as long as 2–3 months. Females can usually move freely between territories and males are unable to coerce them, but in some species such as the northern fur seal, South American sea lion and Australian sea lion, males can successfully contain females in their territories and prevent them from leaving. In some phocid species, like the harbor seal, Weddell seal and bearded seal, the males have underwater territories called "maritories" near female haul-out areas. These are also maintained by vocalizations. The maritories of Weddell seal males can overlap with female breathing holes in the ice.

Male northern elephant seals fighting
for dominance and females.

Lek systems are known to exist among some populations of walruses. These males cluster around females and try to attract them with elaborate courtship displays and vocalizations. Lekking may also exist among California sea lions, South American fur seals, New Zealand sea lions and harbor seals. In some species, including elephant seals and grey seals, males will try to lay claim to the desired females and defend them from rivals. Elephant seal males establish dominance hierarchies with the highest ranking males—the alpha males—maintaining harems of as many as 30–100 females. These males commonly disrupt the copulations of their subordinates while they themselves can mount without inference. They will, however, break off mating to chase off a rival. Grey seal males usually claim a location among a cluster of females whose members may change over time, while males of some walrus populations try to monopolize access to female herds. Male harp seals, crabeater seals and hooded seals follow and defend lactating females in their vicinity—usually one or two at a time, and wait for them to reach estrus.

Younger or subdominant male pinnipeds may attempt to achieve reproductive success in other ways. Subadult elephant seals will sneak into female clusters and try to blend in by pulling in their noses. They also harass and attempt to mate with females that head out to the water. In otariid species like the South American and Australian sea lions, non-territorial subadults form "gangs" and cause chaos within the breeding rookeries to increase their chances of mating with females. Alternative mating strategies also exist in young male grey seals, which do have some success.

Female pinnipeds do appear to have some choice in mates, particularly in lek-breeding species like the walrus, but also in elephant seals where the males try to dominate all the females that they want to mate with. When a female elephant seal or grey seal is mounted by an unwanted male, she tries to squirm and get away, while croaking and slapping him with her tail. This commotion attracts other males to the scene, and the most dominant will end the copulation and attempt to mate with the female himself. Dominant female elephant seals stay in the center of the colony where they are more likely to mate with a dominant male, while peripheral females are more likely to mate with subordinates. Female Steller sea lions are known to solicit mating with their territorial males.

Birth and Parenting

With the exception of the walrus, which has five- to six-year-long inter-birth intervals, female pinnipeds enter estrous shortly after they give birth. All species go through delayed implantation, wherein the embryo remains in suspended development for weeks or months before it is implanted in the uterus. Delayed implantation postpones the birth of young until the female hauls-out on land or until conditions for birthing are favorable. Gestation in seals (including delayed implantation) typically lasts a year. For most species, birthing takes place in the spring and summer months. Typically, single pups are born; twins are uncommon and have high mortality rates. Pups of most species are born precocial.

Harp seal mother nursing pup.

Unlike terrestrial mammals, pinniped milk has little to no lactose. Mother pinnipeds have different strategies for maternal care and lactation. Phocids such as elephant seals, grey seals and hooded seals remain on land or ice and fast during their relatively short lactation period—four days for the hooded seal and five weeks for elephant seals. The milk of these species consist of up to 60% fat, allowing the young to grow fairly quickly. In particular, northern elephant seal pups gain 4 kg (9 lb) each day before they are weaned. Some pups may try to steal extra milk from other nursing mothers and gain weight more quickly than others. Alloparenting occurs in these fasting species; while most northern elephant seal mothers nurse their own pups and reject nursings from alien pups, some do accept alien pups with their own.

Adult Antarctic fur seal with pups.

For otariids and some phocids like the harbor seal, mothers fast and nurse their pups for a few days at a time. In between nursing bouts, the females leave their young onshore to forage at sea. These foraging trips may last anywhere between a day and two weeks, depending on the abundance of food and the distance of foraging sites. While their mothers are away, the pups will fast. Lactation in otariids may last 6–11 months; in the Galápagos fur seal it can last as long as 3 years. Pups of these species are weaned at lower weights than their phocid counterparts. Walruses are unique in that mothers nurse their young at sea. The female rests at the surface with its head held up, and the young suckle upside down. Young pinnipeds typically learn to swim on their own and some species can even swim at birth. Other species may wait days or weeks before entering the water. Elephant seals do not swim until weeks after they are weaned.

Male pinnipeds generally play little role in raising the young. Male walruses may help inexperienced young as they learn to swim, and have even been recorded caring for orphans. Male California sea lions have been observed to help shield swimming pups from predators. Males can also pose threats to the safety of pups. In terrestrially breeding species, pups may get crushed by fighting males. Subadult male South America sea lions sometimes abduct pups from their mothers and treat them like adult males treat females. This helps them gain experience in controlling females. Pups can get severely injured or killed during abductions.

Communication

Walrus males are known to use
vocalizations to attract mates.

Pinnipeds can produce a number of vocalizations such as barks, grunts, rasps, rattles, growls, creaks, warbles, trills, chirps, chugs, clicks and whistles. Vocals are produced both in air and underwater. Otariids are more vocal on land, while phocids are more vocal in water. Antarctic seals are more vocal on land or ice than Arctic seals due to a lack of terrestrial and pagophliic predators like the polar bear. Male vocals are usually of lower frequencies than those of the females. Vocalizations are particularly important during the breeding seasons. Dominant male elephant seals advertise their status and threaten rivals with "clap-threats" and loud drum-like calls that may be modified by the proboscis. Male otariids have strong barks, growls, roars and "whickers". Male walruses are known to produce distinctive gong-like calls when attempting to attract females. They can also create somewhat musical sounds with their inflated throats.

The Weddell seal has perhaps the most elaborate vocal repertoire with separate sounds for airborne and underwater contexts. Underwater vocals include trills, chugs, chirps, chugs and knocks. The calls appear to contain prefixes and suffixes that serve to emphasize a message. The underwater vocals of Weddell seals can last 70 seconds, which is long for a marine mammal call. Some calls have around seven rhythm patterns and are comparable to birdsongs and whalesongs. Similar calls have been recorded in other lobodontine seals and in bearded seals. In some pinniped species, there appear to be geographic differences in vocalizations, known as dialects, while certain species may even have individual variations in expression. These differences are likely important for mothers and pups who need to remain in contact on crowded beaches. Otariid females and their young use mother-pup attraction calls to help them reunite when the mother returns from foraging at sea. The calls are described are "loud" and "bawling". Female elephant seals make an unpulsed attraction call when responding to their young. When threatened by other adults or when pups try to suckle, females make a harsh, pulsed call. Pups may also vocalize when playing, in distress or when prodding their mothers to allow them to suckle. While most vocals are audible to the human ear, a captive leopard seal was recorded making ultrasonic calls underwater. In addition, the vocals of northern elephant seals may produce infrasonic vibrations.

Sea lion balancing a ball.

Non-vocal communication is not as common in pinnipeds as in cetaceans. Nevertheless, when disturbed by intruders harbor seals and Baikal seals may slap their fore-flippers against their bodies as warnings. Teeth chattering, hisses and exhalations are also made as aggressive warnings. Visual displays also occur: Weddell seals will make an S-shaped posture when patrolling under the ice, and Ross seals will display the stripes on their chests and teeth when approached. Male hooded seals use their inflatable nasal membranes to display to and attract females.

Intelligence

In a match-to-sample task study, a single California sea lion was able to demonstrate an understanding of symmetry, transitivity and equivalence; a second seal was unable to complete the tasks. They demonstrate the ability to understand simple syntax and commands when taught an artificial sign language, though they only rarely used the signs semantically or logically. In 2011, a captive California sea lion named Ronan was recorded bobbing its head in synchrony to musical rhythms. This "rhythmic entrainment" was previously seen only in humans, parrots and other birds possessing vocal mimicry. In 1971, a captive harbor seal named Hoover was trained to imitate human words, phrases and laughter. For sea lions used in entertainment, trainers toss a ball at the animal so it may accidentally balance it or hold the ball on its nose, thereby gaining an understanding of the behavior desired. It may require a year to train a sea lion to perform a trick for the public. Its long-term memory allows it to perform a trick after at least three months of non-performance.

In Captivity

Captive sea lion at Kobe Oji Zoo Kobe, Japan.

Pinnipeds can be found in facilities around the world, as their large size and playfulness make them popular attractions. Seals have been kept in captivity since at least Ancient Rome and their trainability was noticed by Pliny the Elder. Zoologist Georges Cuvier noted during the 19th century that wild seals show considerable fondness for humans and stated that they are second only to some monkeys among wild animals in their easily tamability. Francis Galton noted in his landmark paper on domestication that seals were a spectacular example of an animal that would most likely never be domesticated despite their friendliness and desire for comfort due to the fact that they serve no practical use for humans.

Some modern exhibits have rocky backgrounds with artificial haul-out sites and a pool, while others have pens with small rocky, elevated shelters where the animals can dive into their pools. More elaborate exhibits contain deep pools that can be viewed underwater with rock-mimicking cement as haul-out areas. The most common pinniped species kept in captivity is the California sea lion as it is both easy to train and adaptable. Other species popularly kept include the grey seal and harbor seal. Larger animals like walruses and Steller sea lions are much less common. Some organizations, such as the Humane Society of the United States and World Animal Protection, object to keeping pinnipeds and other marine mammals in captivity. They state that the exhibits could not be large enough to house animals that have evolved to be migratory, and a pool could never replace the size and biodiversity of the ocean. They also state that the tricks performed for audiences are "exaggerated variations of their natural behaviors" and distract the people from the animal's unnatural environment. Less entertainment-oriented zoos may still encourage animal play by throwing fish at animals in different directions and providing play equipment.

California sea lions are used in military applications by the U.S. Navy Marine Mammal Program, including detecting naval mines and enemy divers. In the Persian Gulf, the animals have been trained to swim behind divers approaching a U.S. naval ship and attach a clamp with a rope to the diver's leg. Navy officials say that the sea lions can do this in seconds, before the enemy realizes what happened. Organizations like PETA believe that such operations put the animals in danger. The Navy insists that the sea lions are removed once their mission is complete.

Hunting

Men killing northern fur seals on Saint Paul
Island, Alaska, in the mid-1890s.

Humans have hunted seals since the Stone Age. Originally, seals were hit with clubs during haul-out. Eventually, seal hunters used harpoons to spear the animals from boats out at sea, and hooks for killing pups on ice or land. They were also trapped in nets. The use of firearms in seal hunting during the modern era drastically increased the number of killings. Pinnipeds are typically hunted for their meat and blubber. The skins of fur seals and phocids are made into coats, and the tusks of walruses continue to be used for carvings or as ornaments. There is a distinction between the subsistence hunting of seals by indigenous peoples of the Arctic and commercial hunting: subsistence hunters typically use seal products for themselves and depend on them for survival. National and international authorities have given special treatment to aboriginal hunters since their methods of killing are seen as less destructive and wasteful. This distinction is being questioned as indigenous people are using more modern weaponry and mechanized transport to hunt with, and are selling seal products in the marketplace. Some anthropologists argue that the term "subsistence" should also apply to these cash-based exchanges as long as they take place within local production and consumption. More than 100,000 phocids (especially ringed seals) as well as around 10,000 walruses are harvested annually by native hunters.

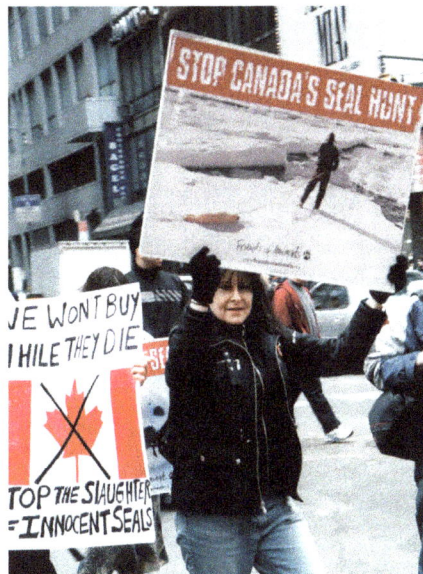
Protests of Canada's seal hunts.

Commercial sealing was historically just as important an industry as whaling. Exploited species included harp seals, hooded seals, Caspian seals, elephant seals, walruses and all species of fur seal. The scale of seal harvesting decreased substantially after the 1960s, after the Canadian government reduced the length of the hunting season and implemented measures to protect adult females. Several species that were commercially exploited have rebounded in numbers; for example, Antarctic fur seals may be as numerous as they were prior to harvesting. The northern elephant seal was hunted to near extinction in the late 19th century, with only a small population remaining on Guadalupe Island. It has since recolonized much of its historic range, but has a population bottleneck. Conversely, the Mediterranean monk seal was extirpated from much of its former range, which stretched from the Mediterranean to the Black Sea and northwest Africa, and only remains in the northeastern Mediterranean and some parts of northwest Africa.

Several species of pinniped continue to be harvested. The Convention for the Conservation of Antarctic Seals allows limited hunting of crabeater seals, leopard seals and Weddell seals. However, Weddell seal hunting is prohibited between September and February if the animal is over one year of age, to ensure breeding stocks are healthy. Other species protected are southern elephant seals, Ross seals and Antarctic fur seals. The Government of Canada permits the hunting of harp seals. This has been met with controversy and debate. Proponents of seal hunts insist that the animals are killed humanely and the white-coated pups are not taken, while opponents argue that it is irresponsible to kill harp seals as they are already threatened by declining habitat.

The Caribbean monk seal has been killed and exploited by Europeans settlers and their descendants since 1494, starting with Christopher Columbus himself. The seals were easy targets for organized sealers, fishermen, turtle hunters and buccaneers because they evolved with little pressure from terrestrial predators and were thus "genetically tame". In the Bahamas, as many as 100 seals were slaughtered in one night. In the mid-nineteenth century, the species was thought to have gone extinct until a small colony was found near the Yucatán Peninsula in 1886. Seal killings continued, and the last reliable report of the animal alive was in 1952. The IUCN declared it extinct in 1996. The Japanese sea lion was common around the Japanese islands, but overexploitation and competition from fisheries drastically decreased the population in the 1930s. The last recorded individual was a juvenile in 1974.

Conservation

As of 2013, the International Union for Conservation of Nature (IUCN) recognizes 35 pinniped species. With the Japanese sea lion and the Caribbean monk seal recently extinct, ten more are considered at risk, as they are ranked "Critically Endangered" (the Mediterranean and Hawaiian monk seals), "Endangered" (Galápagos fur seal, Australian sea lion, Caspian seal and Galápagos sea lion), and "Vulnerable" (northern fur seal, hooded seal and New Zealand sea lion). Three species—the walrus, the ribbon seal, and the spotted seal—have a "Data Deficient" ranking. Species that live in polar habitats are vulnerable to the effects of recent and ongoing climate change, particularly declines in sea ice. There has been some debate over the cause of the decline of Steller sea lions in Alaska since the 1970s.

Some species have become so numerous that they conflict with local people. In the United States, pinnipeds are protected under the Marine Mammal Protection Act of 1972 (MMPA). Since that year, California sea lion populations have risen to 250,000. These animals began exploiting more man-made environments, like docks, for haul-out sites. Many docks are not designed to withstand the weight of several resting sea lions. Wildlife managers have used various methods to control the animals, and some city officials have redesigned docks so they can better withstand use by sea lions. Sea lions also conflict with fisherman since both depend on the same fish stocks. In 2007, MMPA was amended to permit the lethal removal of sea lions from salmon runs at Bonneville Dam. The 2007 law seeks to relieve pressure on the crashing Pacific Northwest salmon populations. Wildlife officials have unsuccessfully attempted to ward off the sea lions using bombs, rubber bullets and bean bags. Efforts to chase sea lions away from the area have also proven ineffective. Critics like the Humane Society object to the killing of the sea lions, claiming that hydroelectric dams pose a greater threat to the salmon. Similar conflicts have existed in South Africa with brown fur seals. In the 1980s and 1990s, South African politicians and fisherman demanded that the fur seals be culled, believing that the animals competed with commercial fisheries. Scientific studies found that culling fur seals would actually have a negative effect on the fishing industry, and the culling option was dropped in 1993.

Grey seal on beach occupied by humans near Niechorze, Poland.
Pinnipeds and humans may compete for space and resources.

Pinnipeds are also threatened by humans indirectly. They are unintentionally caught in fishing nets by commercial fisheries and accidentally swallow fishing hooks. Gillnetting and Seine netting is a significant cause of mortality in seals and other marine mammals. Species commonly entangled include California sea lions, Hawaiian monk seals, northern fur seals and brown fur seals. Pinnipeds are also affected by marine pollution. High levels of organic chemicals accumulate in these animals since they are near the top of food chains and have large reserves of blubber. Lactating mothers can pass the toxins on to their young. These pollutants can cause gastrointestinal cancers, decreased reproductivity and greater vulnerability to infectious diseases. Other man-made threats include habitat destruction by oil and gas exploitation, encroachment by boats, and underwater noise.

SEAL

Seals are web-footed aquatic mammals that live chiefly in cold seas and whose body shape, round at the middle and tapered at the ends, is adapted to swift and graceful swimming. There are two types of seals: the earless, or true, seals (family Phocidae); and the eared seals (family Otariidae), which comprise the sea lions and fur seals. In addition to the presence of external ears, eared seals have longer flippers than do earless seals. Also, the fur of eared seals is more apparent, especially in sea lions.

Seals are carnivores, eating mainly fish, though some also consume squid, other mollusks, and crustaceans. Unlike other seals, the leopard seal (Hydrurga leptonyx) of the Antarctic feeds largely on penguins, seabirds, and other seals, in addition to fish and krill. The main predators of seals are killer whales, polar bears, leopard seals, large sharks, and human beings.

Seal Diversity

The Baikal seal (Phoca sibirica) of Lake Baikal in Siberia, Russia, is the smallest at 1.1–1.4 metres (3.6–4.6 feet) long and 50–130 kg (110–290 pounds), but some female fur seals weigh less. The largest is the male elephant seal (genus Mirounga leonina) of coastal California (including Baja California, Mexico) and South America, which can reach a length of 6.5 metres (21 feet) and a weight of 3,700 kg (8,150 pounds). The upper portions of seals' limbs are within the body, but the long feet and digits remain, having evolved into flippers. Seals possess a thick layer of fat (blubber) below the skin, which provides insulation, acts as a food reserve, and contributes to buoyancy.

Baikal seals (Phoca sibirica), endemic to Lake Baikal,
southeastern Siberia, Russia.

True seals of the genus Phoca are the most abundant in the Northern Hemisphere. They are fairly small, with little difference in size between the sexes. Ringed seals (P. hispida) have blotches over their entire bodies, harp seals (P. groenlandica) have a large blotch of black on otherwise mostly silver-gray fur, harbour seals (P. vitulina) have a marbled coat, and ribbon seals (P. fasciata) have dark fur with ribbons of paler fur around the neck, front limbs, and posterior part of their body.

Though especially abundant in polar seas, seals are found throughout the world, with some species favouring the open ocean and others inhabiting coastal waters or spending time on islands, shores, or ice floes. The coastal species are generally sedentary, but the oceangoing species make extended, regular migrations. All are excellent swimmers and divers—especially the Weddell seal (Leptonychotes weddellii) of the Antarctic. Various species are able to reach depths of 150–250 metres or more and can remain underwater for 20–30 minutes, with the Weddell seal diving for up to 73 minutes and up to 600 metres. Seals cannot swim as fast as dolphins or whales but are more agile in the water. When swimming, a true seal uses its forelimbs to maneuver in the water, propelling its body forward with side-to-side strokes of its hind limbs. Because the hind flippers cannot be moved forward, these seals propel themselves on land by wriggling on their bellies or pulling themselves forward with their front limbs. Eared seals, on the other hand, rely mainly on a rowing motion of their front flippers for propulsion. Because they are able to turn their hind flippers forward, they can use all four limbs when moving on land.

Gray seal (Halichoerus grypus).

All seals must come ashore once a year to breed. Nearly all are gregarious, at least when breeding, with some assembling in enormous herds on beaches or floating ice. Most form pairs during the breeding season, but in some species, such as fur seals, the gray seal (Halichoerus grypus), and elephant seals, males (bulls) take possession of harems of cows and drive rival bulls away from their territory. Gestation periods average about 11 months, including a delayed implantation of the fertilized egg in many species. Cows are again impregnated soon after giving birth. Pups are born on the open ice or in a snow lair on the ice. The mother remains out of the water and does not feed while nursing the pups. The young gain weight rapidly, for the cow's milk is up to about 50 percent fat.

Seals have been hunted for their meat, hides, oil, and fur. The pups of harp seals, for example, are born with white coats that are of value in the fur trade. The fur seals of the North Pacific Ocean and the ringed seals of the North Atlantic Ocean have also been hunted for their pelts. Elephant seals and monk seals were hunted for their blubber, which had various commercial uses. Seal hunting, or sealing, was so widespread and indiscriminate in the 19th century that many species might have become extinct if international regulations had not been enacted for their protection. The severe decline of sealing worldwide after World War II and the effects of international agreements aimed at conserving breeding stocks enabled several severely depleted species to replenish their numbers.

EARED SEAL

Eared seal is the common name for any of the marine mammals comprising the pinniped family Otariidae, characterized by presence of a pinna (external part of ear), the ability to invert their hind-flippers under the body, aiding land movement, and a swimming motion using their long front flippers to propel them through the water. These characteristics help distinguish otariids from the earless seals of the family Phocidae. Extant eared seals comprise 16 species in seven genera commonly known either as sea lions or fur seals.

Otariids are adapted to a semi-aquatic lifestyle, feeding and migrating in the water but breeding and resting on land or ice. They reside in subpolar, temperate, and equatorial waters throughout the Pacific and Southern oceans and the southern Indian and Atlantic oceans. These marine mammals are conspicuously absent in the north Atlantic.

Eared seals play key roles in food chains, consuming fish, mollusks, and sometimes penguins, and being consumed by killer whales, sharks, and bears. They also provide direct values for human beings. Seals have traditionally been hunted for their furs, while sea lions have been trained for such underwater tasks as finding objects or detecting and attaching a clamp to any person underwater who may be approaching military ships or piers.

Eared seals are one of the three main groups of mammals within the taxonomic group Pinnipedia. Pinnipeds are aquatic (mostly marine) mammals that are characterized by having both front and hind limbs in the form of flippers. In addition to eared seals, other pinnipeds are walruses and earless seals.

Eared seals are one of two groups of seals (any pinniped other than walruses): Earless seals, comprising the taxonomic family Phocidae (phocids), and eared seals comprising the family Otariidae (otariids). Walruses generally are considered a separate family of pinnipeds, the Obobenidae, although sometimes they are included with the phocids.

One way of differentiating between the two main groups of seals is by the presence of the pinna, a small furry earflap (external ears), found on the otarids and missing from phocids. Phocids are referred to as "earless seals" because their ears are not easily seen, while otarids are referred to as "eared seals."

In addition to the presence of the pinna, there are other obvious differences between otarids and phocids. Otarids have hind-flippers that can be inverted under the body, aiding their movement on land, while the hind-flippers of phocids cannot be turned forward under the body, causing their movement on land to be slow and awkward. Otarids also swim using their long front flippers to move themselves through the water, while phocids swim by using their rear flippers and lower body in a side-to-side motion. There are also behavioral differences, including the breeding systems.

The eared seals include both fur seals and sea lions. Traditionally, the fur seals were placed in the otariid subfamily Arctocephalinae and the sea lions in the subfamily Otariinae. However, recent studies have suggested that the differences between the fur seals and the sea lions are not great enough to separate them into these two subfamilies.

There are no otariids living in the extreme polar regions; among seals, only earless seals live and breed in the Antarctic and Arctic. On the other hand, a number of fur seals and sea lions live in tropical and subtropical areas, while only two species of phocids (the endangered Hawaiian and Mediterranean monk seals) are found in the tropics and these are small populations; fur seals also range widely into colder climates as well. The Antarctic fur seal (Arctocephalus gazella), which breeds on islands between 45° S and 60° S (95 percent of the population at South Georgia), likely has a winter range that includes spending time close to the Antarctic ice.

Along with the Phocidae and Odobenidae, the two other members of Pinnipedia, Otariidae are considered to be descended from a common ancestor most closely related to modern bears. There remains debate as to whether the phocids diverged from the otariids before or after the walruses.

Otariids arose in the late Miocene (10 to 12 million years ago) in the North Pacific, diversifying rapidly into the Southern Hemisphere, where most species now live. The Callorhinus (northern fur seal) genus is considered to have the oldest lineage.

Traditionally, otariids had been subdivided into the fur seal (Arctocephalinae) and sea lion (Otariinae) subfamilies, with the major distinction between them being the presence of a thick underfur layer in the former. Under this categorization, the fur seals comprised two genera: Callorhinus in the North Pacific with a single representative, the northern fur seal (C. ursinus) and eight species in the southern hemisphere under the genus Arctocephalus, while the sea lions comprise five species under five genera.

Recent analyses of the genetic evidence suggests that the Callorhinus ursinus is in fact more closely related to several sea lion species. Furthermore, many of the Otariinae appear to be more phylogenetically distinct than previously assumed; for example, the Zalophus japonicus is now considered a separate species, rather than a subspecies of Zalophus californius. In light of this evidence, the subfamily separation generally has been removed entirely and the Otariidae family has been organized into seven genera with 16 species and two subspecies.

Nonetheless, because of morphological and behavioral similarity among the "fur seals" and among "sea lions," these remain useful categories when discussing differences between groups of species.

Anatomy and Appearance

Otariids have proportionately much larger fore-flippers and pectoral muscles than phocids, and have the ability to turn their hind limbs forward and walk on all fours, making them far more maneuverable on land. They are generally considered to be less adapted to an aquatic lifestyle, since they breed primarily on land and haul out more frequently than true seals. However, they can attain higher bursts of speed and greater maneuverability in the water. Their swimming power derives from the use of flippers more so than the sinuous whole body movements typical of phocids and walruses.

Otariids are further distinguished by a more dog-like head, sharp, well-developed canines, and the aforementioned visible external pinnae. Their postcanine teeth are generally simple and conical in shape. The dental formula for eared seals is:

3.1.4.1-3

2.1.4.1

In general, fur seals have a more pointed snout and longer fore-flippers than sea lions, and they have a thick, luxuriant coat of fur (pelage). The underfur, which is waterproof, is covered with long, "guard" hairs that give them a "somewhat grizzled appearance". The thick underfur of fur seals have historically made them the objects of commercial exploitation. Sea lions, which generally are larger than fur seals, have a more rounded muzzle and shorter fore-flippers than fur seals, and their pelage is more short and coarse.

Male otariids range in size from the 70 kilograms (150 pounds) Galapagos fur seal, smallest of all pinnipeds, to the over 1000 kilogram (2200 pounds) Steller sea lions. Mature male otariids weigh two to six times more than females with proportionately larger heads, necks, and chests, making them the most sexually dimorphic of all mammals.

Behavior, Feeding and Reproduction

All otariids breed on land during well-defined breeding seasons. Except for the Australian sea lion, which has an atypical 17.5 month breeding cycle, they form strictly annual aggregations on beaches or rocky substrates, often on islands. All species are polygynous; that is, successful males breed with several females.

In most species, males arrive at breeding sites first and establish and maintain territories through vocal and visual displays and occasional fighting. Females typically arrive on shore shortly before giving birth to pups from the mating of the previous year. Females go into estrous sometime after giving birth, perhaps a week or two weeks later, and they breed again, but implanting of the embryo is delayed, allowing an annual cycle in most species.

While considered social animals, there are no permanent hierarchies or statuses established on the colonies. The extent to which males control females or territories varies between species. Northern fur seals and South American sea lions tend to herd specific harem-associated females, occasionally injuring them, while Steller sea lions and New Zealand sea lions control spatial territories but do not generally interfere with the movement of the females.

Otariids are carnivorous, feeding on fish, squid, and krill. Sea lions tend to feed closer to shore in upwelling zones feeding on larger fish while the smaller fur seals tend to take longer, offshore foraging trips and can subsist on large numbers of smaller prey items. They are visual feeders and some females are capable of dives up to 400 meters (1300 feet).

SEA LION

Sea lions are pinnipeds characterized by external ear flaps, long foreflippers, the ability to walk on all fours, short, thick hair, and a big chest and belly. Together with the fur seals, they comprise the family Otariidae, eared seals, which contains six extant and one extinct species (the Japanese sea lion) in five genera. Their range extends from the subarctic to tropical waters of the global ocean in both the Northern and Southern Hemispheres, with the notable exception of the northern Atlantic Ocean. They have an average lifespan of 20–30 years. A male California sea lion weighs on average about 300 kg (660 lb) and is about 2.4 m (8 ft) long, while the female sea lion weighs 100 kg (220 lb) and is 1.8 m (6 ft) long. The largest sea lion is Steller's sea lion, which can weigh 1,000 kg (2,200 lb) and grow to a length of 3.0 m (10 ft). Sea lions consume large quantities of food at a time and are known to eat about 5–8% of their body weight (about 6.8–15.9 kg (15–35 lb)) at a single feeding. Sea lions can go around 16 knots (30 km/h; 18 mph) in water and at their fastest they can go up to 30 knots (56 km/h; 35 mph). Three species, the Australian sea lion, the Galápagos sea lion and the New Zealand sea lion are listed as Endangered.

Taxonomy

Steller sea lions haul out on a rock off the coast of Raspberry Island (Alaska).

Sea lions are related to walruses and seals. Together with the fur seals, they constitute the family Otariidae, collectively known as eared seals. Until recently, sea lions were grouped under a single subfamily called Otariinae, whereas fur seals were grouped in the subfamily Arcocephalinae. This division was based on the most prominent common feature shared by the fur seals and absent in the sea lions, namely the dense underfur characteristic of the former. Recent genetic evidence, suggests Callorhinus, the genus of the northern fur seal, is more closely related to some sea lion species than to the other fur seal genus, Arctocephalus. Therefore, the fur seal/sea lion subfamily distinction has been eliminated from many taxonomies.

Nonetheless, all fur seals have certain features in common: the fur, generally smaller sizes, farther and longer foraging trips, smaller and more abundant prey items, and greater sexual dimorphism. All sea lions have certain features in common, in particular their coarse, short fur, greater bulk, and larger prey than fur seals. For these reasons, the distinction remains useful. The family Otariidae (Order Carnivora) contains the 14 extant species of fur seals and sea lions. Traditional classification of the family into the subfamilies Arctocephalinae (fur seals) and Otariinae (sea lions) is not supported, with the fur seal Callorhinus ursinus having a basal relationship relative to the rest of the family. This is consistent with the fossil record which suggests that this genus diverged from the line leading to the remaining fur seals and sea lions about 6 million years ago (mya). Similar genetic divergences between the sea lion clades as well as between the major Arctocephalus fur seal clades, suggest that these groups underwent periods of rapid radiation at about the time they diverged from each other. The phylogenetic relationships within the family and the genetic distances among some taxa highlight inconsistencies in the current taxonomic classification of the family.

Arctocephalus is characterized by ancestral character states such as dense underfur and the presence of double rooted cheek teeth and is thus thought to represent the most "primitive" line. It was from this basal line that both the sea lions and the remaining fur seal genus, Callorhinus, are thought to have diverged. The fossil record from the western coast of North America presents evidence for the divergence of Callorhinus about 6 mya, whereas fossils in both California and Japan suggest that sea lions did not diverge until years later:

- Suborder Caniformia
  - Family Otariidae
- Subfamily Arctocephalinae
  - Genus Arctocephalus (southern fur seal; eight species)
  - Genus Callorhinus (northern fur seal; one species)
- Subfamily Otariinae
  - Genus Eumetopias
    - Steller's sea lion, E. jubatus
  - Genus Neophoca
    - Australian sea lion, N. cinerea
  - Genus Otaria
    - South American sea lion, O. flavescens
  - Genus Phocarctos
    - New Zealand sea lion or Hooker's sea lion, P. hookeri
  - Genus Zalophus
    - California sea lion, Z. californianus
    - Japanese sea lion, Z. japonicus – extinct (1950s)
    - Galapagos sea lion, Z. wollebaeki
  - Family Phocidae: true seals
  - Family Odobenidae: walrus

Physiology

Diving Adaptations

There are many components that make up sea lion physiology and these processes control aspects of their behavior. Physiology dictates thermoregulation, osmoregulation, reproduction, metabolic rate, and many other aspects on sea lion ecology including but not limited to their ability to dive to great depths. The sea lions' bodies control heart rate, gas exchange, digestion rate, and blood flow to allow individuals to dive for a long period of time and prevent side-effects of high pressure at depth.

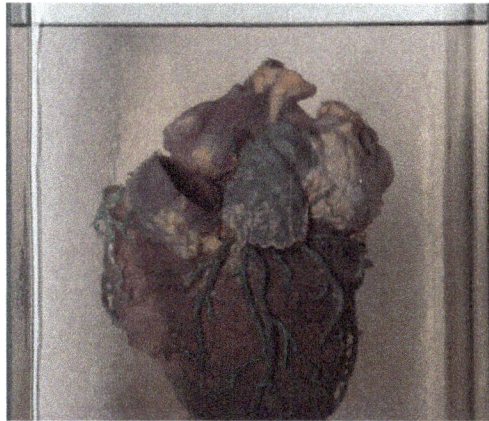

Sea lion heart.

There are many components that make up sea lion physiology and these processes control aspects of their behavior. Physiology dictates thermoregulation, osmoregulation, reproduction, metabolic rate, and many other aspects on sea lion ecology including but not limited to their ability to dive to great depths. The sea lions' bodies control heart rate, gas exchange, digestion rate, and blood flow to allow individuals to dive for a long period of time and prevent side-effects of high pressure at depth.

The high pressures associated with deep dives cause gases such as nitrogen to build up in tissues which are then released upon surfacing, possibly causing death. One of the ways sea lions deal with the extreme pressures is by limiting the amount of gas exchange that occurs when diving. The sea lion allows the alveoli to be compressed by the increasing water pressure thus forcing the surface air into cartilage lined airway just before the gas exchange surface. This process prevents any further oxygen exchange to the blood for muscles, requiring all muscles to be loaded with enough oxygen to last the duration of the dive. However, this shunt reduces the amount of compressed gases from entering tissues therefore reducing the risk of decompression sickness. The collapse of alveoli does not allow for any oxygen storage in the lungs however, this means that sea lions must mitigate oxygen use in order to extend their dives. Oxygen availability is prolonged by the physiological control of heart rate in the sea lions. By reducing heart rate to well below surface rates, oxygen is saved by reducing gas exchange as well as reducing the energy required for a high heart rate. Bradycardia is a control mechanism to allow a switch from pulmonary oxygen to oxygen stored in the muscles which is needed when the sea lions are diving to depth. Another way sea lions mitigate the oxygen obtained at the surface in dives is to reduce digestion rate. Digestion requires metabolic activity and therefore energy and oxygen are consumed during this process, however sea lions can limit digestion rate and decrease it by at least 54%. This reduction in digestion results in a proportional reduction in oxygen use in the stomach and therefore a correlated oxygen supply for diving. Digestion rate in these sea lions increase back to normal rates immediately upon resurfacing. Oxygen depletion limits dive duration, but carbon dioxide (CO_2) build up also plays a role in the dive capabilities of many marine mammals. After a sea lion returns from a long dive, CO_2 is not expired as fast as oxygen is replenished in the blood, due to the unloading complications with CO_2. However, having more than normal levels of CO_2 in the blood does not seem to adversely affect dive behavior. Compared to terrestrial mammals, sea lions have a higher tolerance to storing CO_2 which is what normally tells mammals that they need to breathe. This ability to ignore a response to CO_2 is likely brought on by increase carotid bodies which are sensor for oxygen levels which let the animal know its available oxygen supply. Yet, the sea lions cannot avoid the effects of gradual CO_2 build up which eventually causes the sea lions to spend more time at the surface after multiple repeated dives to allow for enough built up CO_2 to be expired.

Parasites and Diseases

A sea lion at the Memphis zoo.

Behavioural and environmental correlates of Philophthalmus zalophi, a foot parasite. And the infection has impacted the survival of juvenile Galapagos sea lions (Zalophus wollebaeki). This infection leads to diseases that are connected to global warming. The number of infectious stages of different parasites species has a strong correlation with temperature change, therefore it is essential to consider the correlation between the increasing number of parasitic infections and climate changes. To test this proposed theory researchers used Galapagos sea lions because they are endemic to the Galapagos islands. The Galapagos Islands goes through seasonal changes in sea surface temperatures, which consist of high temperatures from the beginning of January through the month of May and lower temperatures throughout the rest of the year. Parasites surfaced in large numbers when the sea temperature was at its highest. Furthermore, data was collected by capturing sea lions in order to measure and determine their growth rates. Their growth rates were noted along with the citings of parasites which were found under the eyelid. The shocking results were that sea lions are affected the parasites from the early ages of 3 weeks old up until the age of 4 to 8 months. The parasites found in the eye fluke did serious damage to the eye. From the data collected, 21 of the 91 survived; with a total of 70 deaths in just a span of two years. The parasites are attacking the pups at such young ages; thus causing the pups to not reach the age of reproduction. The death rates of the pups is surpassing the fertility rate by far. Since most pups are unable to reach the age of reproduction, the population is not growing fast enough to keep the species out of endangerment. The pups who do survive must pass their strong genes down to make sure their young survive and the generation that follows. Other parasites, like Anisakis and heartworm can also infect sea lions.

Along with Galapagos islands, sea lions (Zalophus wollebaeki) being affected are the Australian sea lions (Neophoca cinerea). The same method was used for the sea pups on the galapagos island, but in addition, the researchers in Australia took blood samples. The pups in Australia were being affected by hookworms, but they were also coming out in large numbers with warmer temperatures. Sea pups in New Zealand (Phocarctos hookeri) were also affected really early ages by hookworms (Uncinaria). The difference is that in New Zealand researchers took the necessary steps and began treatment. The treatment seemed to be effective on the pups who have taken it. They found no traces of this infection afterwards. However, the percentage of pups who do have it is still relatively high at about 75%. Those pups who were treated had much better growth rates than those who did not. Overall parasites and hookworms are killing off enough pups to place them in endangerment. Parasites affect sea pups in various areas of the world. Reproductive success reduces immensely, survival methods, changes in health and growth have also been affected.

Gene Expressions and Diet

Diet is an important factor in the well-being of any animal's life. Gene expressions are being used more often to detect the physiological responses to nutrition, as well as other stressors. In a study done with four Steller sea lions (Eumetopias jubatus), three of the four sea lions underwent a 70-day trial which consisted of unrestricted food intake, acute nutritional stress, and chronic nutritional stress. The results of this study showed that the sea lions with nutritional stress down-regulated some cellular processes within their immune response and oxidative stress. Sea lions get affected greatly due to environmental changes because of the dependency they have on marine resources for feeding. A reduced food supply leads to population decline. Compared to many other factors that contribute to an endangered species, nutritional stress is the most proximate cause to population decline.

The New Zealand sea lion has the largest population, therefore no diet studies had ever been conducted. However, when a study was finally conducted the location and climate change effects it had on diet were discovered. North to south composition of a sea lion's diet showed that the temperature gradients were a key factor in the prey mix that was available for the NZ sea lions.

Geographic Variation

The Australian sea lion vs. the Steller sea lion.

Geographic variation for sea lions have been determined by the observations of skulls of several Otariidae species; a general change in size corresponds with a change in latitude and primary productivity. Skulls of Australian sea lions from Western Australia were generally smaller in length whereas the largest skulls are from cool temperate localities. Otariidae are in the process of species divergence, much of which may be driven by local factors, particularly latitude and resources. Populations of a given species tend to be smaller in the tropics, increase in size with increasing latitude, and reach a maximum in sub-polar regions. In a cool climate and cold waters there should be a selective advantage in the relative reduction of body surface area resulting from increased size, since the metabolic rate is related more closely to body surface area than to body weight.

Breeding and Population

Breeding Methods and Habits

Two sea lions on the beach of Otago Peninsula, New Zealand.

Sea lions, with three groups of pinnipeds, have multiple breeding methods and habits over their families but they remain relatively universal. Otariids, or eared sea lions, raise their young, mate, and rest in more earthly land or ice habitats. Their abundance and haul-out behavior have a direct effect on their on land breeding activity. Their seasonal abundance trend correlates with their breeding period between the austral summer of January to March. Their rookeries populate with newborn pups as well as male and female otariids that remain to defend their territories. At the end of the breeding period males disseminate for food and rest while females remain for nurturing. Other points in the year consist of a mix of ages and genders in the rookeries with haul-out patterns varying monthly.

Steller sea lions, living an average of 15 to 20 years, begin their breeding season when adult males establish territories along the rookeries in early May. Male sea lions reach sexual maturity from ages 5 to 7 and don't become territorial until around 9 to 13 years of age. The females arrive in late May bringing in an increase of territorial defense through fighting and boundary displays. After a week births consist most usually of one pup with a perinatal period of 3 to 13 days.

A gathering of more than 40 sea lions off the coast of California.

Steller sea lions have exhibited multiple competitive strategies for reproductive success. Sea lion mating is often polygamous as males usually mate with different females to increase fitness and success, leaving some males to not find a mate at all. Polygamous males rarely provide parental care towards the pup. Strategies used to monopolize females include the resource-defense polygyny, or occupying important female resources. This involves occupying and defending a territory with resources or features attractive to females during sexually receptive periods. Some of these factors may include pupping habitat and access to water. Other techniques include potentially limiting access of other males to females.

Population

A group of sea lions rest in the Ballestas Islands, Peru.

Otaria flavescens (South American sea lion) lives along the Chilean coast with a population estimate of 165,000. According to the most recent surveys in northern and southern Chile the sealing period of the middle twentieth century that left a significant decline in sea lion population is recovering. The recovery is associated with less hunting, otariids rapid population growth, legislation on nature reserves, and new food resources. Haul-out patterns change the abundance of sea lions at particular times of the day, month, and year. Patterns in migration relate to temperature, solar radiation, and prey and water resources. Studies of South American sea lions and other otariids document maximum population on land during early afternoon, potentially due to haul-out during high air temperatures. Adult and subadult males do not show clear annual patterns, maximum abundance being found from October to January. Females and their pups hauled-out during austral winter months of June to September.

Interactions with Humans

GiGi, a sea lion trained by the U.S. Navy for underwater recovery, nuzzles merchant mariner Capt. Arne Willehag of the USNS Sioux during a 1983 training session.

Sea lions entertaining a crowd in Central Park Zoo.

South American sea lions have been greatly impacted by human exploitation. During the late Holocene period to the middle of the twentieth century, hunter gatherers along the Beagle Channel and northern Patagonia had greatly reduced the number of sea lions due to their exploiting and hunting of the species and of the species' environment. Although sealing has been put to a halt in many countries, such as Uruguay, the sea lion population continues to decline because of the drastic effects humans have on their ecosystems. As a result, South American sea lions have been foraging at higher tropical latitudes than they did prior to human exploitation. Fishermen play a key role in the endangerment of sea lions. Sea lions rely on fish, like pollock, as a food source and have to compete with fishermen for it. When fishermen are successful at their job, they greatly reduce the sea lion's food source, which in turn endangers the species. Also, human presence and human recreational activities can cause sea lions to engage in violent and aggressive actions. When humans come closer than 15 meters of a sea lion, the sea lions' vigilance increases because of the disturbance of humans. These disturbances can potentially cause sea lions to have psychological stress responses that cause the sea lions to retreat, sometimes even abandon their locations, and decreases the amount of time sea lions spend hauling out.

Hundreds of California sea lions congregating at Pier 39, San Francisco.

Sea lion attacks on humans are rare, but when humans come within approximately 2.5 meters, it can be very unsafe. In a highly unusual attack in 2007 in Western Australia, a sea lion leapt from the water and seriously mauled a 13-year-old girl surfing behind a speedboat. The sea lion appeared to be preparing for a second attack when the girl was rescued. An Australian marine biologist suggested that the sea lion may have viewed the girl "like a rag doll toy" to be played with. In San Francisco, where an increasingly large population of California sea lions crowds docks along San Francisco Bay, incidents have been reported in recent years of swimmers being bitten on the legs by large, aggressive males, possibly as territorial acts. In April 2015, a sea lion attacked a 62-year-old man who was boating with his wife in San Diego. The attack left the man with a punctured bone. In May 2017, a sea lion dragged a small girl into the water by her dress. She was sitting on a pier side in British Columbia while tourists were illegally feeding the sea lions. She was pulled out of the water with minor injuries and received antibiotic prophylactic treatment for seal finger infection from the superficial bite injury.

Sea lions have also been a focus of tourism in Australia and New Zealand. One of the main sites to view sea lions is in the Carnac Island Nature Reserve near Perth in Western Australia. This tourist site receives over 100,000 visitors, many of whom are recreational boaters and tourists, who can watch the male sea lions haul out on to the shore. They have sometimes been called "the unofficial welcoming committee of the Galápagos Islands".

WALRUS

The walrus (Odobenus rosmarus) is a large flippered marine mammal with a discontinuous distribution about the North Pole in the Arctic Ocean and subarctic seas of the Northern Hemisphere. The walrus is the only living species in the family Odobenidae and genus Odobenus. This species is subdivided into two subspecies: the Atlantic walrus (O. r. rosmarus) which lives in the Atlantic Ocean and the Pacific walrus (O. r. divergens) which lives in the Pacific Ocean.

Adult walrus are easily recognized by their prominent tusks, whiskers, and bulk. Adult males in the Pacific can weigh more than 2,000 kg (4,400 lb) and, among pinnipeds, are exceeded in size only by the two species of elephant seals. Walruses live mostly in shallow waters above the continental shelves, spending significant amounts of their lives on the sea ice looking for benthic bivalve mollusks to eat. Walruses are relatively long-lived, social animals, and they are considered to be a "keystone species" in the Arctic marine regions.

The walrus has played a prominent role in the cultures of many indigenous Arctic peoples, who have hunted the walrus for its meat, fat, skin, tusks, and bone. During the 19th century and the early 20th century, walruses were widely hunted and killed for their blubber, walrus ivory, and meat. The population of walruses dropped rapidly all around the Arctic region. Their population has rebounded somewhat since then, though the populations of Atlantic and Laptev walruses remain fragmented and at low levels compared with the time before human interference.

The walrus is a mammal in the order Carnivora. It is the sole surviving member of the family Odobenidae, one of three lineages in the suborder Pinnipedia along with true seals (Phocidae) and eared seals (Otariidae). While there has been some debate as to whether all three lineages are monophyletic, i.e. descended from a single ancestor, or diphyletic, recent genetic evidence suggests all three descended from a caniform ancestor most closely related to modern bears. Recent multigene analysis indicates the odobenids and otariids diverged from the phocids about 20–26 million years ago, while the odobenids and the otariids separated 15–20 million years ago. Odobenidae was once a highly diverse and widespread family, including at least twenty species in the subfamilies Imagotariinae, Dusignathinae and Odobeninae. The key distinguishing feature was the development of a squirt/suction feeding mechanism; tusks are a later feature specific to Odobeninae, of which the modern walrus is the last remaining (relict) species.

Walrus cows and yearlings (short tusks).

Two subspecies of walrus are widely recognized: the Atlantic walrus, O. r. rosmarus and the Pacific walrus, O. r. divergens. Fixed genetic differences between the Atlantic and Pacific subspecies indicate very restricted gene flow, but relatively recent separation, estimated at 500,000 and 785,000 years ago. These dates coincide with the hypothesis derived from fossils that the walrus evolved from a tropical or subtropical ancestor that became isolated in the Atlantic Ocean and gradually adapted to colder conditions in the Arctic. From there, it presumably recolonized the North Pacific Ocean during high glaciation periods in the Pleistocene via the Central American Seaway.

An isolated population in the Laptev Sea was considered by some authorities, including many Russian biologists and the canonical Mammal Species of the World, to be a third subspecies, O. r. laptevi, but has since been determined to be of Pacific Walrus origin.

Anatomy

Young male Pacific walruses on Cape Pierce in Alaska. Note the variation in the curvature and orientation of the tusks and the bumpy skin (bosses), typical of males.

Walrus using its tusks to hang on a breathing hole in the ice near St. Lawrence Island, Bering Sea.

While some outsized Pacific males can weigh as much as 2,000 kg (4,400 lb), most weigh between 800 and 1,700 kg (1,800 and 3,700 lb). An occasional male of the Pacific subspecies far exceeds normal dimensions. In 1909, a walrus hide weighing 500 kg (1,100 lb) was collected from an enormous bull in Franz Josef Land, while in August 1910, Jack Woodson shot a 4.9 m (16 ft) long walrus, harvesting its 450 kg (1,000 lb) hide. Since a walrus's hide usually accounts for about 20% of its body weight, the total body mass of these two giants is estimated to have been at least 2,300 kg (5,000 lb). The Atlantic subspecies weighs about 10–20% less than the Pacific subspecies. Male Atlantic walrus weigh an average of 900 kg (2,000 lb). The Atlantic walrus also tends to have relatively shorter tusks and somewhat more flattened snout. Females weigh about two-thirds as much as males, with the Atlantic females averaging 560 kg (1,230 lb), sometimes weighing as little as

400 kg (880 lb), and the Pacific female averaging 800 kg (1,800 lb). Length typically ranges from 2.2 to 3.6 m (7.2 to 11.8 ft). Newborn walruses are already quite large, averaging 33 to 85 kg (73 to 187 lb) in weight and 1 to 1.4 m (3.3 to 4.6 ft) in length across both sexes and subspecies. All told, the walrus is the third largest pinniped species, after the two elephant seals. Walruses maintain such a high body weight because of the blubber stored underneath their skin. This blubber keeps them warm and the fat provides energy to the walrus.

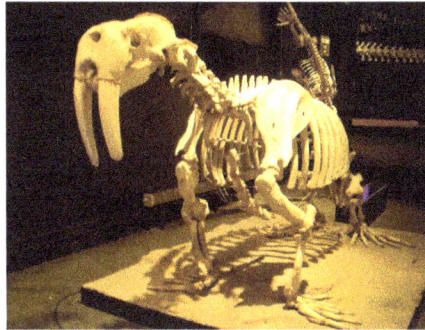
Skeleton.

The walrus's body shape shares features with both sea lions (eared seals: Otariidae) and seals (true seals: Phocidae). As with otariids, it can turn its rear flippers forward and move on all fours; however, its swimming technique is more like that of true seals, relying less on flippers and more on sinuous whole body movements. Also like phocids, it lacks external ears.

The extraocular muscles of the walrus are well-developed. This and its lack of orbital roof allow it to protrude its eyes and see in both a frontal and dorsal direction. However, vision in this species appears to be more suited for short-range.

Tusks and Dentition

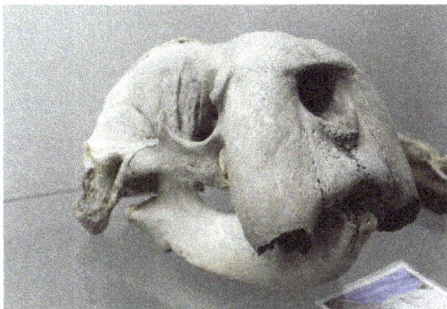
Skull without tusk.

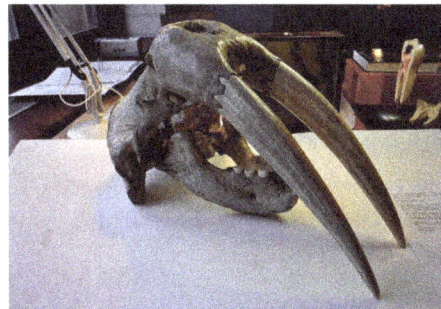
Skull with tusks.

While this was not true of all extinct walruses, the most prominent feature of the living species is its long tusks. These are elongated canines, which are present in both male and female walruses and can reach a length of 1 m (3 ft 3 in) and weigh up to 5.4 kg (12 lb). Tusks are slightly longer and thicker among males, which use them for fighting, dominance and display; the strongest males with the largest tusks typically dominate social groups. Tusks are also used to form and maintain holes in the ice and aid the walrus in climbing out of water onto ice. Tusks were once thought to be used to dig out prey from the seabed, but analyses of abrasion patterns on the tusks indicate they are dragged through the sediment while the upper edge of the snout is used for digging. While the dentition of walruses is highly variable, they generally have relatively few teeth other than the tusks. The maximal number of teeth is 38 with dentition formula: $\frac{3.1.4.2}{3.1.3.2}$, but over half of the teeth are rudimentary and occur with less than 50% frequency, such that a typical dentition includes only 18 teeth $\frac{1.1.3.0}{0.1.3.0}$.

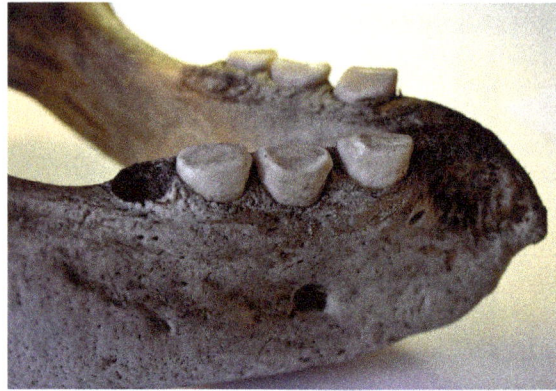

Tooth.

Vibrissae

Surrounding the tusks is a broad mat of stiff bristles ('mystacial vibrissae'), giving the walrus a characteristic whiskered appearance. There can be 400 to 700 vibrissae in 13 to 15 rows reaching 30 cm (12 in) in length, though in the wild they are often worn to much shorter lengths due to constant use in foraging. The vibrissae are attached to muscles and are supplied with blood and nerves, making them highly sensitive organs capable of differentiating shapes 3 mm (0.12 in) thick and 2 mm (0.079 in) wide.

Skin

Aside from the vibrissae, the walrus is sparsely covered with fur and appears bald. Its skin is highly wrinkled and thick, up to 10 cm (3.9 in) around the neck and shoulders of males. The blubber layer beneath is up to 15 cm (5.9 in) thick. Young walruses are deep brown and grow paler and more cinnamon-colored as they age. Old males, in particular, become nearly pink. Because skin blood vessels constrict in cold water, the walrus can appear almost white when swimming. As a secondary sexual characteristic, males also acquire significant nodules, called "bosses", particularly around the neck and shoulders.

The walrus has an air sac under its throat which acts like a flotation bubble and allows it to bob vertically in the water and sleep. The males possess a large baculum (penis bone), up to 63 cm (25 in) in length, the largest of any land mammal, both in absolute size and relative to body size.

Reproduction

Walruses live to about 20–30 years old in the wild. The males reach sexual maturity as early as seven years, but do not typically mate until fully developed at around 15 years of age. They rut from January through April, decreasing their food intake dramatically. The females begin ovulating as soon as four to six years old. The females are diestrous, coming into heat in late summer and also around February, yet the males are fertile only around February; the potential fertility of this second period is unknown. Breeding occurs from January to March, peaking in February. Males aggregate in the water around ice-bound groups of estrous females and engage in competitive vocal displays. The females join them and copulate in the water.

Walruses fighting.

A herd of walruses on Northbrook Island, Franz Josef Land, Russia.

Gestation lasts 15 to 16 months. The first three to four months are spent with the blastula in suspended development before it implants itself in the uterus. This strategy of delayed implantation, common among pinnipeds, presumably evolved to optimize both the mating season and the birthing season, determined by ecological conditions that promote newborn survival. Calves are born during the spring migration, from April to June. They weigh 45 to 75 kg (99 to 165 lb) at birth and are able to swim. The mothers nurse for over a year before weaning, but the young can spend up to five years with the mothers. Walrus milk contains higher amounts of fats and protein compared to land animals but lower compared to phocid seals. This lower fat content in turn causes a slower growth rate among calves and a longer nursing investment for their mothers. Because ovulation is suppressed until the calf is weaned, females give birth at most every two years, leaving the walrus with the lowest reproductive rate of any pinniped.

A walrus pup at Kamogawa Seaworld, Japan.

Migration

The rest of the year (late summer and fall), walruses tend to form massive aggregations of tens of thousands of individuals on rocky beaches or outcrops. The migration between the ice and the beach can be long-distance and dramatic. In late spring and summer, for example, several hundred thousand Pacific walruses migrate from the Bering Sea into the Chukchi Sea through the relatively narrow Bering Strait.

Ecology

Range and Habitat

The majority of the population of the Pacific walrus spends its summers north of the Bering Strait in the Chukchi Sea of the Arctic Ocean along the northern coast of eastern Siberia, around Wrangel Island, in the Beaufort Sea along the north shore of Alaska south to Unimak Island, and in the waters between those locations. Smaller numbers of males summer in the Gulf of Anadyr on the southern coast of the Siberian Chukchi Peninsula, and in Bristol Bay off the southern coast of Alaska, west of the Alaska Peninsula. In the spring and fall, walruses congregate throughout the Bering Strait, reaching from the western coast of Alaska to the Gulf of Anadyr. They winter over in the Bering Sea along the eastern coast of Siberia south to the northern part of the Kamchatka Peninsula, and along the southern coast of Alaska. A 28,000-year-old fossil walrus was dredged up from the bottom of San Francisco Bay, indicating Pacific walruses ranged that far south during the last ice age.

Commercial harvesting reduced the population of Pacific walrus to between 50,000-100,000 in the 1950s-1960s. Limits on commercial hunting allowed the population to increase to a peak in the 1970s-1980s, but subsequently, walrus numbers have again declined. Early aerial censuses of Pacific walrus conducted at five-year intervals between 1975 and 1985 estimated populations of above 220,000 in each of the three surveys. In 2006, the population of Pacific walrus was estimated to be around 129,000 on the basis of an aerial census combined with satellite tracking. There were roughly 200,000 Pacific walruses in 1990.

The much smaller population of Atlantic walruses ranges from the Canadian Arctic, across Greenland, Svalbard, and the western part of Arctic Russia. There are eight hypothetical subpopulations of walruses, based largely on their geographical distribution and movements: five west of Greenland and three east of Greenland. The Atlantic walrus once ranged south to Sable Island, Nova Scotia, and as late as the eighteenth century was found in large numbers in the greater Gulf of St. Lawrence region, sometimes in colonies of up to 7,000 to 8,000 individuals. This population was nearly eradicated by commercial harvest; their current numbers, though difficult to estimate, probably remain below 20,000. In April 2006, the Canadian Species at Risk Act listed the population of the northwest Atlantic walrus in Quebec, New Brunswick, Nova Scotia, Newfoundland and Labrador as having been eradicated in Canada.

The isolated population of Laptev walruses is confined year-round to the central and western regions of the Laptev Sea, the eastmost regions of the Kara Sea, and the westmost regions of the East Siberian Sea. The current population of these walruses has been estimated to be between 5,000 and 10,000. The limited diving abilities of walruses brings them to depend on shallow waters (and the nearby ice floes) for reaching their food supply.

Diet

Walruses prefer shallow shelf regions and forage primarily on the sea floor, often from sea ice platforms. They are not particularly deep divers compared to other pinnipeds; their deepest recorded dives are around 80 m (260 ft). They can remain submerged for as long as half an hour.

Vibrissae of captive walrus (Japan).

The walrus has a diverse and opportunistic diet, feeding on more than 60 genera of marine organisms, including shrimp, crabs, tube worms, soft corals, tunicates, sea cucumbers, various mollusks, and even parts of other pinnipeds.] However, it prefers benthic bivalve mollusks, especially clams, for which it forages by grazing along the sea bottom, searching and identifying prey with its sensitive vibrissae and clearing the murky bottoms with jets of water and active flipper movements. The walrus sucks the meat out by sealing its powerful lips to the organism and withdrawing its piston-like tongue rapidly into its mouth, creating a vacuum. The walrus palate is uniquely vaulted, enabling effective suction. The diet of the Pacific walrus consist almost exclusively of benthic invertebrates (97 percent).

Walruses leaving the water.

Aside from the large numbers of organisms actually consumed by the walrus, its foraging has a large peripheral impact on benthic communities. It disturbs (bioturbates) the sea floor, releasing nutrients into the water column, encouraging mixing and movement of many organisms and increasing the patchiness of the benthos.

Seal tissue has been observed in fairly significant proportion of walrus stomachs in the Pacific, but the importance of seals in the walrus diet is under debate. There have been isolated observations of walruses preying on seals up to the size of a 200 kg (440 lb) bearded seal. Rarely, incidents of walruses preying on seabirds, particularly the Brünnich's guillemot (Uria lomvia), have been documented. Walruses may occasionally prey on ice-entrapped narwhals and scavenge on whale carcasses but there is little evidence to prove this.

Predators

Due to its great size and tusks, the walrus has only two natural predators: the killer whale (orca) and the polar bear. The walrus does not, however, comprise a significant component of either of these predators' diets. Both the orca and the polar bear are also most likely to prey on walrus calves. The polar bear often hunts the walrus by rushing at beached aggregations and consuming the individuals crushed or wounded in the sudden exodus, typically younger or infirm animals. The bears also isolate walruses when they overwinter and are unable to escape a charging bear due to inaccessible diving holes in the ice. However, even an injured walrus is a formidable opponent for a polar bear, and direct attacks are rare. Walruses have been known to fatally injure polar bears in battles if the latter follows the other into the water where the bear is at a disadvantage. Polar bear–walrus battles are often extremely protracted and exhausting, and bears have been known to forgo the attack after injuring a walrus. Orcas regularly attack walrus, although walruses are believed to have successfully defended themselves via counterattack against the larger cetacean. However, orcas have been observed successfully attacking walruses with few or no injuries.

Conservation

Trained walrus in captivity at Marineland.

 In the 18th and 19th centuries, the walrus was heavily exploited by American and European sealers and whalers, leading to the near extirpation of the Atlantic population. Commercial walrus harvesting is now outlawed throughout its range, although Chukchi, Yupik and Inuit peoples are permitted to kill small numbers towards the end of each summer.

Traditional hunters used all parts of the walrus. The meat, often preserved, is an important winter nutrition source; the flippers are fermented and stored as a delicacy until spring; tusks and bone were historically used for tools, as well as material for handicrafts; the oil was rendered for warmth and light; the tough hide made rope and house and boat coverings; and the intestines and gut linings made waterproof parkas. While some of these uses have faded with access to alternative technologies, walrus meat remains an important part of local diets, and tusk carving and engraving remain a vital art form.

According to Adolf Erik Nordenskiöld, European hunters and Arctic explorers found walrus meat not particularly tasty, and only ate it in case of necessity; however walrus tongue was a delicacy.

Walrus hunts are regulated by resource managers in Russia, the United States, Canada, and Denmark, and representatives of the respective hunting communities. An estimated four to seven thousand Pacific walruses are harvested in Alaska and in Russia, including a significant portion (about 42%) of struck and lost animals. Several hundred are removed annually around Greenland. The sustainability of these levels of harvest is difficult to determine given uncertain population estimates and parameters such as fecundity and mortality. The Boone and Crockett Big Game Record book has entries for Atlantic and Pacific walrus. The recorded largest tusks are just over 30 inches and 37 inches long respectively.

The effects of global climate change are another element of concern. The extent and thickness of the pack ice has reached unusually low levels in several recent years. The walrus relies on this ice while giving birth and aggregating in the reproductive period. Thinner pack ice over the Bering Sea has reduced the amount of resting habitat near optimal feeding grounds. This more widely separates lactating females from their calves, increasing nutritional stress for the young and lower reproductive rates. Reduced coastal sea ice has also been implicated in the increase of stampeding deaths crowding the shorelines of the Chukchi Sea between eastern Russia and western Alaska. However, there are insufficient climate data to make reliable predictions on population trends.

Currently, two of the three walrus subspecies are listed as "least-concern" by the IUCN, while the third is "data deficient". The Pacific walrus is not listed as "depleted" according to the Marine Mammal Protection Act nor as "threatened" or "endangered" under the Endangered Species Act. The Russian Atlantic and Laptev Sea populations are classified as Category 2 (decreasing) and Category 3 (rare) in the Russian Red Book. Global trade in walrus ivory is restricted according to a CITES Appendix 3 listing. In October 2017, the Center for Biological Diversity announced they would sue the U.S. Fish and Wildlife Service to force it to classify the Pacific Walrus as a threatened or endangered species.

References

- Fish, F. E. (2003). "Maneuverability by the sea lion Zalophus californianus: Turning performance of an unstable body design". Journal of Experimental Biology. 206 (4): 667–74. Doi:10.1242/jeb.00144. PMID 12517984

- Pinniped, animal: britannica.com, Retrieved 9 February, 2019

- Etymology of mammal names. Iberianature.com (29 December 2010). Retrieved 16 September 2011.

- Seal-mammal, animal: britannica.com, Retrieved 10 March, 2019

- Chilvers, B. L. (2015). "Phocarctos hookeri. The IUCN Red List of Threatened Species". International Union for Conservation of Nature and Natural Resources. Doi:10.2305/IUCN.UK.2015-2.RLTS.T17026A1306343.en

- Eared-seal, entry: newworldencyclopedia.org, Retrieved 11 April, 2019

- Tegiminis (20 November 2014). "Why Sealioning Is Bad". Simplikation.com. Retrieved 31 October 2016

- Born, E. W.; Andersen, L. W.; Gjertz, I. & Wiig, Ø (2001). "A review of the genetic relationships of Atlantic walrus (Odobenus rosmarus rosmarus) east and west of Greenland". Polar Biology. 24 (10): 713–718. Doi:10.1007/s003000100277

5

Sirenians

Sirenians are an order of fully aquatic, herbivores mammals that live in rivers, swamps, estuaries and marine wetlands. Some of the different types of sirenians are West Indian manatees, African manatees and dugongs. This chapter discusses in detail these types of sirenians and their characteristics.

Sirenia is a small order, composed of just two extant families, Dugongidae and Trichechidae, with four current species. Family Trichechidae includes three species: West Indian manatees (Trichechus manatus), African manatees (Trichechus senegalensis), and Amazonian manatees (Trichechus inunguis). There is only one extant member of family Dugongidae, dugongs (Dugong dugon). Dugongs are more streamlined than manatees, they lack nails on their flippers, and have a bi-lobed tail. Steller's sea cows (Hydrodamalis gigas) are a recently extinct dugong species, and they were the only sirenians that did not inhabit tropical waters, instead they were found in the subarctic Bering Sea. Steller's sea cows represent the largest known members of Sirenia; growing up to 10 meters and weighing up to 11,000 kg; whereas the smallest known members, little sea cows, weighed approximately 150 kg, extant sirenians often weigh between 400 to 1,500 kg. Steller's sea cows were also unique due to their lack of teeth; instead, they had keratinized masticatory plates on the inside of their mouth, which they used to grind their food. The order name Sirenia is based on sirens, also known as sea nymphs, as the mermaid myths likely originate with these animals.

Geographic Range

All extant sirenians are found in shallow waters along coastlines and inlets. Manatees are found along tropical coastlines on both sides of the Atlantic Ocean and in the Amazon Basin. Dugongs are found off of coastal eastern Africa, along the shores of the Indian Ocean, and on the northern coast of Australia. The extinct Steller's sea cows were found in coastal waters of the Bering Sea.

Habitat

Sirenians inhabit a variety of tropical and subtropical aquatic habitats. All prefer water at least two meters deep, with an abundance of submerged, aquatic vegetation. Sirenians primarily inhabit coastal, marine habitats, but Amazonian manatees inhabit strictly freshwater habitats in the Amazon Basin. Sirenians in subtropical areas inhabit warm, fresh water during colder months, and are often seen in the warm water near coastal power plants. In the warmer months, they move to tepid saline waters. Sirenians are unable to tolerate water temperatures below a certain threshold. West Indian manatees migrate if the water temperature falls below 20°C (68°F), African manatees prefer temperatures above 18 °C (64 °F), Amazonian manatees prefer higher temperatures, about 25° to 30 °C (77° to 86 °F), and dugongs tolerate temperatures no lower than 19 °C (66 °F). If temperatures drop below these levels it can be fatal. In fact, one of the largest mortality factors for sirenians is exposure to cold waters. This is especially true for West Indian manatees.

Physical Description

Sirenians are large, slow-moving, aquatic mammals. They are torpedo-shaped with long, broad backs tapering to paddle-like, dorso-ventrally flattened tails; the tail is spoon-shaped in family Trichechidae and dugongs have bi-lobed tails. Sirenians have two flippers; manatees have three to four nails on the second, third and fourth digits, while dugongs lack nails. All sirenians lack hind limbs, and have gray-brown skin that is smooth in some species, such as Amazonian manatees, or wrinkled in others, such as West Indian manatees. Adults are between 2.8 and 3.5 meters long. African and West Indian manatees weigh between 1,000 and 1,500 kg. Dugongs and Amazonian manatees weigh much less, around 400 kg. Sirenians' dense pachystotic bones, along with their long, thin lungs, help them overcome buoyancy issues. Females have one teat and are often heavier than males. The rostrum is deflected downwards, reflecting their preference for submerged, aquatic vegetation. This feature is especially exaggerated in dugongs, which are strictly bottom feeders. Around the mouth there are many short bristles. Sirenians lack external pinnae. They have a pair of semi-circular valves on their nose, which they close during diving and open during surface breathing. Their teeth are low-crowned and bicuspid. These teeth are lost throughout their lifetime due to their coarse diet. Teeth are replaced from the back of the mouth, with 5 to 7 teeth in the upper and lower jaw at any given time. Through the course of their lifetime, sirenians can go through up to 30 teeth per jaw quadrant.

Reproduction

Sirenians are polygamous and have an approximately 1 to 1 sex ratio. When a cow undergoes estrus, she attracts herds of males who follow her while she swims to evade them. Females may use this evasive swimming behavior to select a superior mate. Males who happen upon others engaging in sexual behavior are often encouraged to participate. These groups of males may remain near a single female for several weeks before a successful copulation. During this time, males and females often grasp each other with their flippers as a form of sexual play. Males may show aggression toward each other; male dugongs often exhibit scars from the tusks of other males. Prior to mating, the male reorients the female with his flippers; they mate ventral to ventral, which may require moving into deeper water. Two notable exceptions to this mating system are dugongs in Shark Bay, Australia and Steller's sea cows. Male dugongs in Shark Bay defend territories and attempt to attract females, rather than actively pursuing them. Steller's sea cows may have been monogamous, living with long lasting family groups consisting of a mating pair and two offspring. Steller's sea cows appeared to show strong mate fidelity and were even observed to remain near the body of a dead mate for days.

Female sirenians are believed to be polyestrous, although the length of their estrous cycle is not known. Sirenian gestation lasts 12 to 14 months. Cows travel into shallow waters to give birth; occasionally females enter water so shallow they are nearly beached and must wait for the tide to float out again. All sirenians are capable of breeding year round, but each species has seasonal peaks when birth rates are particularly high. For instance, Amazonian manatees have more young during the wet season when food is abundant. African manatees also have a higher occurrence of birth during the wet season, while dugongs and West Indian manatees tend to have higher birth rates during warmer months. Females typically give birth to one precocial calf, although about 1.8% of manatee births result in twins. The calf may need assistance reaching the surface to take its first breath, but swims on its own before the end of its first day. Sirenians reach sexual maturity in 3 to 10 years and reproduce every 2.5 to 7 years; cows may breed more frequently if they lose a very young calf.

Among extant sirenian species, the only strong social bonds are formed between the mother and calf. Calves are born with teeth and begin foraging in their third month, they generally continue nursing for about 18 months, although they may nurse up to four years. Very young calves synchronize all activities, such as breathing and resting, to match their mother. Young calves often cling to their mother and ride on her back. If they feel their calf is in danger, a female may face and even headbutt a predator, although much more often the cow and calf flee together. Historically, dugongs have gathered in large herds, several hundred strong. Although contemporary herds still occur, they are much smaller and form less often. These herds are believed to form so calves can learn to swim in a protected environment. There is evidence that male Steller's sea cows provided some parental care by defended calves from predation; there are even reports of bulls attacking boats to protect their young.

Lifespan/Longevity

The average sirenian lifespan ranges from 50 to 70 years, the maximum lifespan is estimated to be around 73 years. Causes of adult mortality include predation, habitat disturbance, hunting, poisoning from various pollutants, and illness, especially in cooler waters. Recently, illness has become a serious issue due to over-crowding from habitat loss. It is estimated that sirenians must achieve a survival rate of 90% or higher to maintain their population size. Manatees fare much better in captivity than dugongs, which are often captured at especially young ages and face longer transportation routes.

Behavior

In general, sirenians are neither nocturnal nor diurnal; instead they are equally active all day. However, some populations particularly susceptible to hunting have become more active at night. Sirenians display agitated behavior if they sense a nearby predator and often make impulsive chirping sounds. Groups of sirenians may even stand against and headbutt predators, but more typically, they attempt to flee danger. While feeding, sirenians often use their flippers to "walk" along the sea bottom. Sirenians beat their tails up and down to achieve forward motion. When accelerating, sirenians bring their flippers up against their body, but at cruising speed, the flippers typically hang down. Flippers are used to aid in turning and stopping. Although they can achieve speeds of 25 km/hr when being pursued, sirenians typically move about 3 to 10 km/hr.

Sirenians do not travel along any set migratory paths, but do exhibit some seasonality in their travels. Amazonian manatees spend the dry season in deep lakes or river channels, although they prefer to spend the wet seasons in flooded forests where food is more abundant. During this period, Amazonian manatees are less active and subsist on a reduced diet. There is evidence that African manatees exhibit similar behavioral patterns during the African dry season. Dugongs and West Indian manatees also travel into tropical waters if temperatures begin to dip during the winter, although West Indian manatees have been known to congregate around springs, power plants, or other sources of warm water instead. Sirenians may also exhibit daily travel patterns. Their nature is fairly nomadic; they may spend days, weeks, or even seasons in one spot, or they may travel tens of kilometers in a single day. Sirenians travel to deeper waters to avoid rough weather. Because of their dependence on seagrasses, it is not necessary for sirenians to dive to great depths. They cruise 1 to 3 m below the surface, the greatest recorded depth for a dive is 10 m. Sirenians typically come to the surface to breathe every 2 to 4 minutes, the greatest recorded interval between breaths is 18 minutes.

Communication and Perception

Sirenians' communicate by sound; this communication is best developed between a mother and calf. Cows and calves use vocalizations to keep track of one another, it is believed that these animals can identify and distinguish one another based on their chirps and barks. Mothers may respond to their calf from over 50 m. Sirenians also produce sounds when fearful, sexually aroused, or playful. Male dugongs produce low frequency barks when competing for a mate. All calls are usually very short in duration. Sirenians can hear over a frequency range of 0.4 to 46 kHz, but their peak hearing range is 6 to 20 kHz. Sirenians' visual capabilities are comparable to that of humans. The visual field of their eyes overlap and sirenians blink often to keep their eyes lubricated. Sirenians may also have a well-developed tactile sense; they have sensory hairs all over their body and on their rostrum, and a touch-sensitive epidermis. Sirenians touch each other frequently during sexual play, and often rub against rocks during leisure. Little is known about the gustation and olfaction of sirenians, although they are believed to have an extremely poor sense of smell.

Food Habits

Sirenians are herbivorous, eating mainly plants such as seagrasses, water weeds, and other aquatic vegetation. Manatees tend to be opportunistic feeders, eating a wide variety of plant matter, including mangrove leaves and Hydrilla, as well as leaves and acorns from overhanging branches on the bank, they have even been known to eat floating palm fruit. While manatees consume at least 60 different plant species, dugongs are more particular; they are strictly bottom feeders and prefer seagrass. This specialized diet is evident in their highly inflected rostrum. The extinct Steller's sea cow is thought to have fed primarily on kelp. Sirenians are occasionally known to consume clams or fish, possibly for their protein. They also tend to ingest some small invertebrates that reside on the plants they eat; however, although the protein content is ultimately beneficial, their consumption is likely unintentional. It unclear how sirenians obtain the fresh water their bodies need. Some manatees live in freshwater, and those that live in saltwater are often attracted to freshwater areas, such as river mouths or water pipes. Dugongs, however, live strictly in saltwater. It is suspected that the kidneys of sirenians, especially dugongs, may have a special ability to filter salt water, otherwise, they may obtain all of their needed freshwater from the plants they eat.

Each day, sirenians spend up to 8 hours feeding and eat about 5 to 10% of their body weight. They do not seem to have any specific time preference, as sirenians appear to have both nocturnal and diurnal feeding times. Sirenians often use the thick pads on their upper and lower jaw to help tear or bite their food. Dugongs in particular use their flippers to pull seagrasses out of the seafloor, eating the roots and rhizomes as well. Because dugongs remove the entire plant, it is easy to follow the trail they leave along the ocean floor. Sirenians are constantly moving in search of food, and do not appear to be territorial of feeding sites. Rather, sirenians simply move where food is available. These animals use hind-gut digestion, and food takes about 7 days to travel through the digestive system. Their metabolic rate is only 36% of what is predicted by their body size. Due to lack of food, Amazonian manatees often fast during the dry season; they have been known to survive up to 200 days without food. However, that length of food deprivation could be fatal, especially because they have known to consume soil or clays in desperation, which may also kill them. When food is sparse, breeding may also be delayed.

Predation

Sirenians' main predators are humans. Some cultures hunt sirenians for sustenance or spiritual reasons, but these animals most often fall prey to unintentional predators such as nets, barges, and flood control gates. Sirenians have a few natural predators in certain regions. For instance, Amazonian manatees are occasionally preyed on by jaguars, caimans, and sharks, dugongs may be attacked by tiger sharks, crocodiles, and killer whales. In the case of tiger sharks, dugongs simply avoid the predator by feeding elsewhere, even if the seagrasses in the secondary location are sub-par. However, any attacks seem to be isolated, and for the most part, sirenians have no significant natural predators.

Ecosystem Roles

Sirenians often host a number of small internal and external parasites. Common parasites include nematodes, flukes, and tapeworms. Many fluke species infect the intestines, nasal passages, or lungs. Sirenians also host a number of commensals which inhabit their skin, including copepods, barnacles, remoras, diatoms, and algae. These creatures do not harm sirenians and are probably of little significance to their overall health. For example, sharksuckers (Echeneis naucrates and Echeneis neucratoides) attach themselves to sirenians to feed on their fecal matter.

Sirenians impact their ecosystem through their herbivory on aquatic vegetation (up to 90 kg/day), which can be especially significant during temporary winter aggregations in areas with warmer water. The paleoecology of sirenians has been extensively studied, especially with regard to the Miocene and Oligocene when they were more diverse. Fossil records show a direct relationship between sirenian abundance and sea grass and kelp abundance. It is thought that sirenian feeding habits could have permanently altered aquatic landscapes of the past.

Commensal/Parasitic Species:

- Nematodes (phylum Nematoda).

- Flukes (phylum Platyhelminthes; class Trematoda).

- Tapeworms (phylum Platyhelminthes; class Cestoda).

- Copepods (phylum Arthropoda; class Maxillopoda; subclass Copepoda).

- Barnacles (phylum Arthropoda; class Maxillopoda; infraclass Cirripedia).

- Common remoras (Remora remora).

- Sharksuckers (Echeneis naucrates).

- Whitefin sharksuckers (Echeneis neucratoides).

Economic Importance for Humans

Positive

In some parts of the world, such as New Guinea and west Africa, manatees continue to be hunted for their meat, hide, and oil, which are often sold. However, in most parts of the world, sirenians are protected under law. Living sirenians could potentially serve other useful purposes that would economically benefit humans. For instance, manatees could be used as an inexpensive method of weed control in problem areas. In tests of this theory, the results have varied, the weeds usually disappear, but sometimes the manatees die as well. Likewise, they could help control mosquito infestations, as removing excess aquatic vegetation may also reduce mosquito populations.

Negative

There are no known negative impacts of sirenians on humans. Some have suggested they harm fish populations or fishing gear, but this is unsubstantiated.

Conservation Status

All extant sirenians are considered vulnerable according to the IUCN. Populations of all species are in decline due to hunting and injuries associated with boat impacts. There are many extinct sirenian species. Four subfamilies of Dugongidae are now extinct, including species such as Steller's sea cows (Hydrodamalis gigas), which likely went extinct around 1768 due to over-hunting. There are many laws to protect these creatures but they are often incompletely enforced. There are extensive programs to protect West Indian manatees in the United States and Australia has also made strides in dugong preservation by establishing reserves and research facilities.

MANATEE

Manatees (family Trichechidae, genus Trichechus) are large, fully aquatic, mostly herbivorous marine mammals sometimes known as sea cows. There are three accepted living species of Trichechidae, representing three of the four living species in the order Sirenia: the Amazonian manatee (Trichechus inunguis), the West Indian manatee (Trichechus manatus), and the West African manatee (Trichechus senegalensis). They measure up to 4.0 metres (13.1 ft) long, weigh as much as 590 kilograms (1,300 lb), and have paddle-like flippers. Manatees are occasionally called sea cows, as they are slow plant-eaters, peaceful and similar to cows on land. They often graze on water plants in tropical seas.

Manatees weigh 400 to 550 kilograms (880 to 1,210 lb), and average 2.8 to 3.0 metres (9.2 to 9.8 ft) in length, sometimes growing to 4.6 metres (15 ft) and 1,775 kilograms (3,913 lb) (the females tend to be larger and heavier). At birth, baby manatees weigh about 30 kilograms (66 lb) each. The manatee has a large, flexible, prehensile upper lip, used to gather food and eat and for social interaction and communication. Manatees have shorter snouts than their fellow sirenians, the dugongs. The lids of manatees' small, widely spaced eyes close in a circular manner. The adults have no incisor or canine teeth, just a set of cheek teeth, which are not clearly differentiated into molars and premolars. These teeth are repeatedly replaced throughout life, with new teeth growing at the rear as older teeth fall out from farther forward in the mouth, somewhat as elephants' teeth do. At any time, a manatee typically has no more than six teeth in each jaw of its mouth. Its tail is paddle-shaped, and is the clearest visible difference between manatees and dugongs; a dugong tail is fluked, similar in shape to that of a whale. The female manatee has two teats, one under each flipper, a characteristic that was used to make early links between the manatee and elephants.

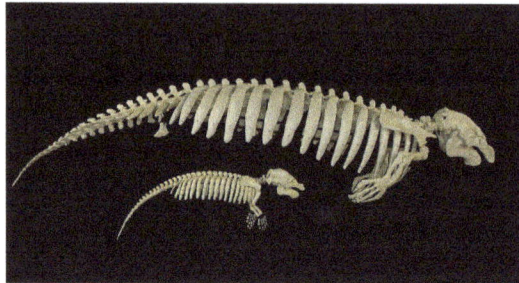

A skeleton of a manatee and calf, on display at The Museum
of Osteology, Oklahoma City, Oklahoma.

The manatee is unusual among mammals in having just six cervical vertebrae, a number that may be due to mutations in the homeotic genes. All other mammals have seven cervical vertebrae, other than the two-toed and three-toed sloths.

Skull of a West Indian manatee on display at The
Museum of Osteology, Oklahoma City, Oklahoma.

Like the horse, the manatee has a simple stomach, but a large cecum, in which it can digest tough plant matter. Generally, the intestines are about 45 meters, unusually long for an animal of the manatee's size.

Taxonomy

Manatees are three of the four living species in the order Sirenia. The fourth is the Eastern Hemisphere's dugong. The Sirenia are thought to have evolved from four-legged land mammals more than 60 million years ago, with the closest living relatives being the Proboscidea (elephants) and Hyracoidea (hyraxes).

The Amazonian's hair color is brownish gray, and it has thick wrinkled skin, often with coarse hair, or "whiskers". Photos are rare; although very little is known about this species, scientists think it is similar to West Indian manatee.

Behavior

Endangered Florida manatee (Trichechus manatus).

Apart from mothers with their young, or males following a receptive female, manatees are generally solitary animals. Manatees spend approximately 50% of the day sleeping submerged, surfacing for air regularly at intervals of less than 20 minutes. The remainder of the time is mostly spent grazing in shallow waters at depths of 1–2 metres (3.3–6.6 ft). The Florida subspecies (T. m. latirostris) has been known to live up to 60 years.

Locomotion

Generally, manatees swim at about 5 to 8 kilometres per hour (3 to 5 mph). However, they have been known to swim at up to 30 kilometres per hour (20 mph) in short bursts.

Intelligence and Learning

Manatee postures in captivity.

Manatees are capable of understanding discrimination tasks and show signs of complex associative learning. They also have good long-term memory. They demonstrate discrimination and task-learning abilities similar to dolphins and pinnipeds in acoustic and visual studies.

Reproduction

Manatees typically breed once every two years; generally only a single calf is born. Gestation lasts about 12 months and to wean the calf takes a further 12 to 18 months.

Communication

Manatees emit a wide range of sounds used in communication, especially between cows and their calves. Their ears are large internally but the external openings are small, and they are located four inches behind each eye. Adults communicate to maintain contact and during sexual and play behaviors. Taste and smell, in addition to sight, sound, and touch, may also be forms of communication.

Diet

Manatees are herbivores and eat over 60 different freshwater (e.g., floating hyacinth, pickerel weed, alligator weed, water lettuce, hydrilla, water celery, musk grass, mangrove leaves) and saltwater plants (e.g., sea grasses, shoal grass, manatee grass, turtle grass, widgeon grass, sea clover, and marine algae). Using their divided upper lip, an adult manatee will commonly eat up to 10%–15% of their body weight (about 50 kg) per day. Consuming such an amount requires the manatee to graze for up to seven hours a day. To be able to cope with the high levels of cellulose in their plant based diet, manatees utilize hindgut fermentation to help with the digestion process. Manatees have been known to eat small numbers of fish from nets.

Feeding Behavior

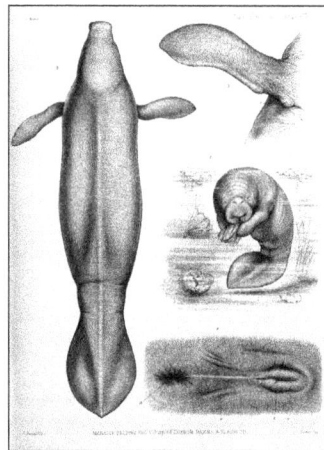

Manatee plate.

Manatees use their flippers to "walk" along the bottom whilst they dig for plants and roots in the substrate. When plants are detected, the flippers are used to scoop the vegetation toward the manatee's lips. The manatee has prehensile lips; the upper lip pad is split into left and right sides which can move independently. The lips use seven muscles to manipulate and tear at plants. Manatees use their lips and front flippers to move the plants into the mouth. The manatee does not have front teeth, however, behind the lips, on the roof of the mouth, there are dense, ridged pads. These horny ridges, and the manatee's lower jaw, tear through ingested plant material.

Dentition

Manatees have four rows of teeth. There are 6 to 8 high-crowned, open-rooted molars located along each side of the upper and lower jaw giving a total of 24 to 32 flat, rough-textured teeth. Eating gritty vegetation abrades the teeth, particularly the enamel crown; however, research indicates that the enamel structure in manatee molars is weak. To compensate for this, manatee teeth are continually replaced. When anterior molars wear down, they are shed. Posterior molars erupt at the back of the row and slowly move forward to replace these like enamel crowns on a conveyor belt, similarly to elephants. This process continues throughout the manatee's lifetime. The rate at which the teeth migrate forward depends on how quickly the anterior teeth abrade. Some studies indicate that the rate is about 1 cm/month although other studies indicate 0.1 cm/month.

Ecology

Range and Habitat

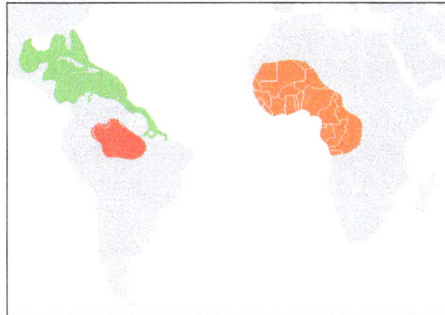

Approximate distribution of Trichechus; T. manatus in green;
T. inunguis in red; T. senegalenis in orange.

Manatees inhabit the shallow, marshy coastal areas and rivers of the Caribbean Sea and the Gulf of Mexico (T. manatus, West Indian manatee), the Amazon basin (T. inunguis, Amazonian manatee), and West Africa (T. senegalensis, West African manatee).

West Indian manatees prefer warmer temperatures and are known to congregate in shallow waters. They frequently migrate through brackish water estuaries to freshwater springs. They cannot survive below 15 °C (60 °F). Their natural source for warmth during winter is warm, spring-fed rivers.

A group of three manatees.

West Indian

The coast of the state of Georgia is usually the northernmost range of the West Indian manatees because their low metabolic rate does not protect them in cold water. Prolonged exposure to water below 20 °C (68 °F) can cause "cold stress syndrome" and death. Florida manatees can move freely between fresh water and salt water.

Manatees have been seen as far north as Cape Cod, and in 1995 and again in 2006, one was seen in New York City and Rhode Island's Narragansett Bay. A manatee was spotted in the Wolf River harbor near the Mississippi River in downtown Memphis in 2006, and was later found dead 10 miles downriver in McKellar Lake.

The West Indian manatee migrates into Florida rivers — such as the Crystal, the Homosassa, and the Chassahowitzka rivers, whose headsprings are 22 °C (72 °F) all year. In November to March, about 400 West Indian manatees (according to the National Wildlife Refuge) gather in the rivers in Citrus County, Florida.

In winter, manatees often gather near the warm-water outflows of power plants along the Florida coast, instead of migrating south as they once did. Some conservationists are concerned that these manatees have become too reliant on these artificially warmed areas. The U.S. Fish and Wildlife Service is trying to find a new way to heat the water for manatees that depended on plants that have closed. The main water treatment plant in Guyana has four manatees that keep storage canals clear of weeds; there are also some in the ponds of the national park in Georgetown, Guyana.

Accurate population estimates of the Florida manatee (T. manatus) are difficult. They have been called scientifically weak because they vary widely from year to year, some areas showing increases, others decreases, and little strong evidence of increases except in two areas. Manatee counts are highly variable without an accurate way to estimate numbers: In Florida in 1996, a winter survey found 2,639 manatees; in 1997, a January survey found 2,229, and a February survey found 1,706. A statewide synoptic survey in January 2010 found 5,067 manatees living in Florida, the highest number recorded to that time.

As of January 2016, the USFWS estimates the range-wide manatee population to be at least 13,000; as of January, 2018, at least 6,100 are estimated to be in Florida.

Population viability studies conducted in 1997 found that decreasing adult survival and eventual extinction were a probable future outcome for Florida manatees unless they got more protection. The U.S. Fish and Wildlife Service proposed downgrading the manatee's status from endangered to threatened in January 2016 after more than 40 years of the manatee's being classified as on the endangered. Fossil remains of Florida manatee ancestors date back about 45 million years.

Amazonian

The freshwater Amazonian manatee (T. inunguis) inhabits the Amazon River and its tributaries, and never ventures into salt water.

West African

They are found in coastal marine and estuarine habitats, and in freshwater river systems along the west coast of Africa from the Senegal River south to the Cuanza River in Angola. They live as far upriver on the Niger River as Koulikoro in Mali, 2,000 km from the coast.

Predation

Overall, predation does not present a significant threat to the survival of any manatee species.

Threats

Young manatees can be curious; this individual is inspecting a kayak.

Antillean manatee.

The main causes of death for manatees are human-related issues, such as habitat destruction and human objects. Natural causes of death include adverse temperatures, predation by crocodiles on young, and disease.

Ship Strikes

Their slow-moving, curious nature, coupled with dense coastal development, has led to many violent collisions with propeller-driven boats and ships, leading frequently to maiming, disfigurement, and even death. As a result, a large proportion of manatees exhibit spiral cutting propeller scars on their backs, usually caused by larger vessels that do not have skegs in front of the propellers like the smaller outboard and inboard-outboard recreational boats have. They are now even identified by humans based on their scar patterns. Many manatees have been cut in two by large vessels like ships and tug boats, even in the highly populated lower St. Johns River's narrow channels. Some are concerned that the current situation is inhumane, with upwards of 50 scars and disfigurements from vessel strikes on a single manatee. Often, the lacerations lead to infections, which can prove fatal. Internal injuries stemming from being trapped between hulls and docks and impacts have also been fatal. Recent testing shows that manatees may be able to hear speed boats and other watercraft approaching, due to the frequency the boat makes. However, a manatee may not be able to hear the approaching boats when they are performing day-to-day activities or distractions. The manatee has a tested frequency range of 8 kilohertz to 32 kilohertz.

Manatees hear on a higher frequency than would be expected for such large marine mammals. Many large boats emit very low frequencies, which confuse the manatee and explain their lack of awareness around boats. The Lloyd's mirror effect results in low frequency propeller sounds not being discernible near the surface, where most accidents occur. Research indicates that when a boat has a higher frequency the manatees rapidly swim away from danger.

In 2003, a population model was released by the United States Geological Survey that predicted an extremely grave situation confronting the manatee in both the Southwest and Atlantic regions where the vast majority of manatees are found. It states,

> In the absence of any new management action, that is, if boat mortality rates continue to increase at the rates observed since 1992, the situation in the Atlantic and Southwest regions is dire, with no chance of meeting recovery criteria within 100 years.

> "Hurricanes, cold stress, red tide poisoning and a variety of other maladies threaten manatees, but by far their greatest danger is from watercraft strikes, which account for about a quarter of Florida manatee deaths," said study curator John Jett.

Manatee bearing scars on its back from a boat propeller.

According to marine mammal veterinarians:

- The severity of mutilations for some of these individuals can be astounding – including long term survivors with completely severed tails, major tail mutilations, and multiple disfiguring dorsal lacerations. These injuries not only cause gruesome wounds, but may also impact population processes by reducing calf production (and survival) in wounded females – observations also speak to the likely pain and suffering endured. In an example, they cited one case study of a small calf "with a severe dorsal mutilation trailing a decomposing piece of dermis and muscle as it continued to accompany and nurse from its mother by age 2 its dorsum was grossly deformed and included a large protruding rib fragment visible."

These veterinarians go on to state:

- The overwhelming documentation of gruesome wounding of manatees leaves no room for denial. Minimization of this injury is explicit in the Recovery Plan, several state statutes, and federal laws, and implicit in our society's ethical and moral standards.

In 2009, of the 429 Florida manatees recorded dead, 97 were killed by commercial and recreational vessels, which broke the earlier record number of 95 set in 2002.

Red Tide

Another cause of manatee deaths are red tides, a term used for the proliferation, or "blooms", of the microscopic marine algae, Karenia brevis. This dinoflagellate produces brevetoxins that can have toxic effects on the central nervous system of animals.

In 1996, a red tide was responsible for 151 manatee deaths. The bloom was present from early March to the end of April and killed approximately 15% of the known population of manatees along South Florida's western coast. Other blooms in 1982 and 2005 resulted in 37 and 44 deaths, respectively.

Additional Threats

Manatees can also be crushed and isolated in water control structures (navigation locks, floodgates, etc.) and are occasionally killed by entanglement in fishing gear, such as crab pot float lines, box traps, and shark nets.

While humans are allowed to swim with manatees in one area of Florida, there have been numerous charges of people harassing and disturbing the manatees. According to the United States Fish and Wildlife Service, approximately 99 manatee deaths each year are related to human activities. In January 2016, there were 43 manatee deaths in Florida alone.

Conservation

All three species of manatee are listed by the World Conservation Union as vulnerable to extinction. It is illegal under federal and Florida law to injure or harm a manatee. They are classified as "endangered" by both the state and the federal governments.

The MV Freedom Star and MV Liberty Star, ships used by NASA to tow space shuttle solid rocket boosters back to Kennedy Space Center, are propelled only by water jets to protect the endangered manatee population that inhabits regions of the Banana River where the ships are based.

Brazil outlawed hunting in 1973 in an effort to preserve the species. Deaths by boat strikes are still common. In January 2016, the U.S. Fish and Wildlife Service proposed that the West Indian manatee be reclassified from an "endangered" status to "threatened" as improvements to habitat conditions, population growth and reductions of threats have all increased. The proposal will not affect current federal protections. As of February 2016, 6,250 manatees were reported swimming in Florida's springs.

Hunting

Manatees were traditionally hunted by indigenous Caribbean people. When Christopher Columbus arrived in the region, hunting was already an established trade, although this is less common today.

Trichechus sp.

The primary hunting method was for the hunter to approach in a dugout canoe, offering bait to attract it close enough to temporarily stun it with a blow near the head from an oar-like pole. Many times the creature would flip over, leaving it vulnerable to further attacks.

From manatee hides, Native Americans made war shields, canoes, and shoes, though manatees were predominantly hunted for their abundant meat.

Later, manatees were hunted for their bones, which were used to make "special potions". Until the 1800s, museums paid as much as $100 for bones or hides. Though hunting was banned in 1893, poaching continues today.

Captivity

A manatee at SeaWorld, Florida.

The oldest manatee in captivity was Snooty, at the South Florida Museum's Parker Manatee Aquarium in Bradenton, Florida. Born at the Miami Aquarium and Tackle Company on July 21, 1948, Snooty was one of the first recorded captive manatee births. Raised entirely in captivity, Snooty was never to be released into the wild. As such he was the only manatee at the aquarium, and one of only a few captive manatees in the United States that was allowed to interact with human handlers. That made him uniquely suitable for manatee research and education.

Snooty died suddenly two days after his 69th birthday, July 23, 2017, when he was found in an underwater area only used to access plumbing for the exhibit life support system. The South Florida Museum's initial press release stated, "Early indications are that an access panel door that is normally bolted shut had somehow been knocked loose and that Snooty was able to swim in."

There are a number of manatee rehabilitation centers in the United States. These include three government-run critical care facilities in Florida at Lowry Park Zoo, Miami Seaquarium, and SeaWorld Orlando. After initial treatment at these facilities, the manatees are transferred to rehabilitation facilities before release. These include the Cincinnati Zoo and Botanical Garden, Columbus Zoo and Aquarium, Epcot's The Seas, South Florida Museum, and Homosassa Springs Wildlife State Park.

The Columbus Zoo was a founding member of the Manatee Rehabilitation Partnership in 2001. Since 1999, the zoo's Manatee Bay facility has helped rehabilitate 20 manatees. The Cincinnati Zoo has rehabilitated and released more than a dozen manatees since 1999.

Manatees can also be viewed in a number of European zoos, such as the Tierpark Berlin, the Nuremberg Zoo, in ZooParc de Beauval in France and in the Aquarium of Genoa in Italy. The River Safari at Singapore features seven of them. They are also included in the plans of the Wild Place Project in Bristol, England, whose first exhibit is opened in summer 2013 with the manatees as an addition as early as 2015.

WEST INDIAN MANATEE

The West Indian manatee (Trichechus manatus) or "sea cow", also known as American manatee, is the largest surviving member of the aquatic mammal order Sirenia (which also includes the dugong and the extinct Steller's sea cow).

The West Indian manatee is a species distinct from the Amazonian manatee (T. inunguis) and the African manatee (T. senegalensis). Based on genetic and morphological studies, the West Indian manatee is divided into two subspecies, the Florida manatee (T. m. latirostris) and the Antillean or Caribbean manatee (T. m. manatus). However, recent genetic (mtDNA) research suggests that the West Indian manatee actually consists of three groups, which are more or less geographically distributed as: (1) Florida and the Greater Antilles; (2) Mexico, Central America and northern South America; and (3) northeastern South America.

The West Indian Manatee was placed on the Endangered Species List in the 1970s, when there were only several hundred left, and it has been of great conservation concern to federal, state, private, and nonprofit organizations to protect these species from natural and human-induced threats like collisions with boats. On March 30, 2017, the US Secretary of the Interior Ryan Zinke announced the federal reclassification of the manatee from endangered to threatened as the number of sea cows had increased to over 6,000.

Like the other sirenians, the West Indian manatee has adapted fully to aquatic life, having no hind limbs. Pelage cover is sparsely distributed across the body, which may play a role in reducing the build-up of algae on their thick skin. The average West Indian manatee is about 2.7–3.5 m (8.9–11.5 ft) long and weighs 200–600 kg (440–1,320 lb), with females generally larger than males. The difference between the two subspecies of the West Indian manatee is that the Florida manatee is commonly reported as being larger in size compared to Antillean manatee. The largest individual on record weighed 1,655 kg (3,649 lb) and measured 4.6 m (15 ft) long. This manatee's color is gray or brown. Its flippers also have either three or four nails.

Distribution and Habitat

As its name implies, the West Indian manatee lives in the West Indies, or Caribbean, generally in shallow coastal areas. However, it is known to withstand large changes in water salinity, so has also been found in shallow rivers and estuaries. It can live in fresh, brackish, and saline water. It is limited to the tropics and subtropics due to an extremely low metabolic rate and lack of a thick layer of insulating body fat. While this is a regularly occurring species along coastal southern Florida, during summer, this large mammal has even been found as far north as Dennis, Massachusetts, and as far west as Texas. A manatee was spotted in the Wolf River (near where it enters the Mississippi) in Memphis, Tennessee in 2006.

Manatee from Crystal River, Florida.

The Florida manatee (Trichechus manatus latirostris), a subspecies of the West Indian manatee, is the largest of all living sirenians. Florida manatees inhabit the most northern limit of sirenian habitats. Over three decades of research by universities, governmental agencies, and NGOs have contributed to understanding of Florida manatee ecology and behavior. They are found in freshwater rivers, in estuaries, and in the coastal waters of the Gulf of Mexico and the Atlantic Ocean. Florida manatees may live to be more than 28 years old in the wild, and one captive manatee, "Snooty", lived for 69 years.

Large concentrations of Florida manatees are located in the Crystal River and Blue Springs regions in central and north Florida, as well as along the Atlantic Coast, and Florida Gulf Coast.

The other subspecies of the West Indian manatee is sometimes referred to as the Antillean manatee (T. m. manatus). Antillean manatees are sparsely distributed throughout the Caribbean and the northwestern Atlantic Ocean, from Mexico, east to the Greater Antilles, and south to Brazil. They are found in The Bahamas, French Guiana, Suriname, Guyana, Trinidad, Venezuela, Colombia, Panama, Costa Rica, Nicaragua, Honduras, Guatemala, Belize, Mexico, Cuba, Haiti, the Dominican Republic, Jamaica, and Puerto Rico. Historically, Antillean manatees were hunted by local natives and sold to European explorers for food. Today, they are threatened by loss of habitat, poaching, entanglement with fishing gear, and vessel strikes.

Behavior and Diet

Behavior

Basking at Haulover Canal, Merritt Island
National Wildlife Refuge, Florida.

The West Indian manatee is surprisingly agile in water, and individuals have been seen doing rolls, somersaults, and even swimming upside-down. Manatees are not territorial and do not have complex predator avoidance behavior, as they have evolved in areas without natural predators. The common predators of marine mammals, such as killer whales and large sharks, are rarely (if ever) found in habitats inhabited by this species.

Communication

Based upon their behavior, Bauer et al. suggest that manatees may share the characteristic of pheromonal communication with their relative, the elephant. Some scientists have observed that manatees form long periods of mating herds when wandering males come across estrous females, which indicates the possibility that males are able to sense the estrogen or other chemical indicators. Other scientists have observed that manatees can communicate information to each other through their vocalization patterns. Evidence suggests that there are sex and age-related differences in the vocalization structure of common squeaks and screeches in adult males, adult females, and juveniles. This may be an indication of vocal individuality among manatees. An increase in Manatee vocalization after a vocal playback stimulus shows that they may be able to recognize another Manatee's individual voice. This behavior in manatees is found mostly between mother and calf interactions. However, vocalization can still be commonly found in a variety of social interactions within groups of manatees, which is similar to other aquatic mammals. When communicating in noisy environments, manatees that are in groups experience the same Lombard effect as humans do; where they will involuntary increase their vocal effort when communicating in loud environments. Based on acoustic and anatomical evidence, mammalian vocal folds are assumed to be the mechanism for sound production in manatees. Manatees also eat other manatees' feces; it is assumed that they do this to gather information about reproductive status or dominance indicating the important role chemoreception plays in the social and reproductive behavior of manatees.

Diet

Manatees are obligate herbivores that feed on over 60 species of aquatic plants in both fresh and salt water. In addition, when the tide is high enough, they will also feed on grasses and leaves. They also consume some fish and small invertebrates. While many manatees are known to eat a large quantity throughout the day, the amount they eat depends on their body size and activity level. Manatees typically graze for 5 or more hours per day consuming anywhere from 4% to 10% of their body weight in wet vegetation per day. Because manatees feed on abrasive plants, their molars are often worn down and are replaced many times throughout their lives, so they are called "marching molars". The molar teeth are similar in shape, but of varying sizes. Replacement of the molar teeth are done so in the forward direction. Manatees do not have incisors. In fact, the incisors have been replaced by horny gingival plates.

Manatees are nonruminats with an enlarged hindgut. Unlike other hindgut fermenters, such as the horse, manatees efficiently extract nutrients, particularly cellulose, from the aquatic plants in their diet. Manatees have a large gastrointestinal tract with contents measuring about 23% of its total body mass. In addition, the passage rate of food is very long (about 7 days). Having an increased rate of digestion is beneficial by increasing the digestibility of their diet. It is suggested that chronic fermentation may also provide additional heat and is correlated with their low metabolic rate.

Vibrissae

Sculpture of manatee showing vibrissae.

All the hairs of the manatee may be vibrissae.

Manatees have sensitive tactile hairs that cover their bodies and faces called vibrissae. Each individual hair is a vibrissal apparatus known as a follicle-sinus complex. Vibrissae are blood filled sinuses bound by a dense connective tissue capsule with sensitive nerve endings that provides haptic feedback to the manatee.

Usually vibrissae are found on the facial regions of terrestrial and non-sirenian aquatic animals and are called whiskers. Manatees, however, have vibrissae all over their bodies. The vibrissae located in their facial region are roughly 30 times denser than the vibrissae on the rest of their body.

Their mouth consists of very mobile prehensile lips which are used for grasping food and objects. The vibrissae on these lips are turned outward during grasping and are used in locating vegetation. Their oral disks also contain vibrissae which have been classified as bristle-like hairs that are used in nongrasping investigation of objects and food.

Research has found that manatee vibrissae are so sensitive that they are able to perform active touch discrimination of textures. Manatees also use their vibrissae to navigate the turbid waterways of their environment. Research has indicated that they are able to use these vibrissae to detect hydrodynamic stimuli in the same way that fish use their lateral line system.

Reproduction

Although female West Indian manatees are mostly solitary creatures, they form mating herds while in estrus. Most females first breed successfully between ages of seven and nine; they are, however, capable of reproduction as early as four years of age. Most males reach sexual maturity by the time they are three or four. The gestation period is 12 to 14 months. Normally, one calf is born, although on rare occasions two have been recorded. The young are born with molars, allowing them to consume sea grass within the first three weeks of birth. On average, manatees that survive to adulthood will have between five and seven offspring between the ages of 20 and 26.When a calf is born, it usually weighs 60–70 lb (27–32 kg) and is 4.0–4.5 ft (1.2–1.4 m) long. The family unit consists of mother and calf, which remain together for up to two years. Males aggregate in mating herds around a female when she is ready to mate, but contribute no parental care to the calf.

Manatee Conservation

Manatees in a conservation project in
Brazilian northeastern coast.

The West Indian manatee has been hunted for hundreds of years for meat and hide, and continues to be hunted in Central and South America. Illegal poaching, as well as collisions with vessels, are a constant source of manatee fatalities. Additionally, environmental stresses such as red tide and cold waters cause several health problems to manatees such as immunosuppression, disease, and even death.

The Florida manatee subspecies (T. m. latirostris) was listed in October 2007 as endangered by the International Union for the Conservation of Nature (IUCN) on the basis of a population size of less than 2,500 mature individuals and a population estimated to be in decline by at least 20% over the next two generations (estimated at about 40 years) due to anticipated future changes in warm-water habitat and threats from increasing watercraft traffic over the next several decades.

According to the US Fish & Wildlife Service (FWS) and the Florida Fish & Wildlife Conservation Commission, the IUCN "endangered" category is equivalent to the US Endangered Species Act (ESA) category of "threatened." In 2013 the manatee was listed under the ESA as "endangered," which is equivalent to the IUCN category of "critically endangered." In April 2007, the US Fish and Wildlife Service advised the species be reclassified as threatened rather than endangered. The Florida Fish & Wildlife Conservation Commission no longer includes the manatee on its list of state imperiled species.

In 2007, the Florida Manatee Biological Review Panel presented their assessment of the Florida Manatee for the year 2005-2006. They reported that there were no "statistically-based estimates (with variance) of abundance for the entire Florida manatee population" and that the highest count obtained from surveys undertaken since 1991, was "3300 manatees in January 2001." The 2007 Biological Review Panel assessment confirmed that the greatest threat to the Florida manatee population was the potential future loss of warm-water habitat. The West Indian manatee has a high casualty rate due to thermal shock from cold temperatures. During cold weather, many die due to their digestive tracts shutting down at water temperatures below 20 °C (68 °F). The Florida manatee is a tropical species unable to tolerate water temperatures below 20 °C (68 °F). During the winter months, over 300 manatees often congregate near the warm water outflows of power plants along the coast of Florida instead of migrating south as they once did, causing some conservationists to worry that manatees have become too reliant on these artificially warmed areas. According to the Florida Fish and Wildlife Conservations Commission (2010), a recorded 237 manatees died that year with 42% of those fatalities being a result of cold stress syndrome. The US Fish and Wildlife Service is trying to find a new way to heat the water for manatees that are dependent on plants that have closed. According to Alvarez-Aleman et al. (2010), the first known Florida manatee was recorded utilizing the warm waters expelled by a power plant canal in Cuba in July 2006 and the following year in January, February, and April, a mother manatee and her calf were reported at the power plant in Havana, Cuba.

Many manatee deaths are caused by both large and small boats. Manatees are also at a disadvantage because they are not able to quickly move away from an oncoming boat.

Thirty-eight percent of manatee deaths, between the years 1995 and 2005, were caused by human-induced activities such as boats, water control devices, fishing equipment, and toxic chemicals; therefore, conservation strategies involving effective public education programs and public policy enforcement are useful to manage these anthropogenic-induced fatal tragedies. Researchers strongly suggest that manatees' oral temperature, heart rate, and respiration rate should be strongly monitored during all human interventions such as field research, rescue, and captivity. Additionally, since studies have shown that death does not appear to be a common result of capture, it is believed that capture and care is necessary for manatees inhabiting Florida, Puerto Rico, and Belize. One conservation strategy in maintaining viable population size is manatee rehabilitation. According to The Society for Conservation Biology the four goals of manatee conservation include conservation science, conservation management, education, and policy.

Manatee veterinarian with the FWC's Fish and Wildlife Research Institute, "We know that watercraft-related mortality is still the main threat to manatees long-term". By 2007, watercraft collisions accounted for about 25% of all documented manatee deaths and was "the single greatest known cause of mortality" and the greatest limiting factor to the speed at which the manatee population could recover from stochastic events. In 2016, 104 manatee deaths were water-craft related. In the same year, 520 dead manatees were found in waterways across Florida, "the third deadliest year since record-keeping began" according to Florida Fish and Wildlife Conservation Commission.

West Indian manatee skeletons on display at the North Carolina
Museum of Natural Sciences in Raleigh, North Carolina.

Agencies responsible for administering the US Endangered Species Act are obliged to provide updates to the Manatee Core Biological Model every five years. The 2013 the authors of "Manatee Core Biological Model" concluded that in Florida, "statewide, the likelihood of a 50% or greater decline in three manatee generations was 12%; the likelihood of a 20% or greater decline in two generations was 56%.": The Core Biological Model and a related manatee threats analysis, prepared by the US Geological Service for the US Fish & Wildlife Service, represented a significant improvement in data collection and analysis. The models project that the "estimated probability that the statewide population will fall below 1000 animals within 100 years was 2.3%."

According to the Pacific Legal Foundation (PLF), USFWS failed to follow-up on its 2007 report by Haubold et al. by drafting a downlisting proposal for the manatee from endangered to threatened. On behalf of Save Crystal River, PLF petitioned the FWS to downlist the manatee. The species were reclassified federally by the US Fish and Wildlife Service on March 30, 2017 from endangered to threatened as the number of sea cows had increased from a few hundred in the 1970s to over 6,000 in the 2010s. According to Save the Manatee Club the USFWS decision failed to adequately consider data from 2010 to 2016, during which time manatees suffered from unprecedented mortality events linked to habitat pollution, dependence on artificial warm water sources, and record deaths from watercraft strikes.

USFWS announced that the manatee is still protected under the Marine Mammal Protection Act and that all federal protections for the manatee remain in place. Until being removed from the endangered list, Manatees received protection from the US Endangered Species Act of 1973. The West Indian manatee is still protected by the Florida Manatee Sanctuary Act of 1978 and the US Marine Mammal Protection Act of 1972.

Sonar technology is also aiding in the West Indian Manatee conservation. Side Scan technology was first introduced in the 1960s, and within the past 20 years, it has been used to aid in manatee conservation. There are currently four ways side scan is being used to research and aid in conservation. These are to aid in detection other than visually, determine group sizes and mother-calf relationships, determining habitat preferences and detection, and assisting in manatee captures. It has been shown that this technology produces a much more accurate detection percentage (>80%) than simply visual detection. Future implications include better estimates of populations and population dynamics.

AFRICAN MANATEE

The African manatee (Trichechus senegalensis), also known as the West African manatee or sea cow, is a species of manatee that is mostly herbivorous. African manatees inhabit much of the western region of Africa – from Senegal to Angola. Although not a great deal is known about this species, it is hypothesized that the African manatee is very similar to the West Indian manatee (Trichechus manatus).

The African manatee's body is widest at the middle, and its tail resembles a paddle. The manatee is gray in color with small, colorless hairs that cover its body. However, algae and other tiny organisms often grow on an African manatee's body, so its body sometimes appears brown or greenish in color. Calves are darker in color when they are very young. African manatees measure up to 4.5 m (15 ft) in length, and weigh about 360 kilograms (790 pounds). African manatees are typically extremely slow, moving between 4.8 km and 8.0 km (3 and 5 mi) per hour, although when scared by predators they can travel at speeds of about 32 km (20 mi) per hour. The African manatee's large forelimbs, or flippers, are used to paddle and to bring food to the its mouth. Vegetation is then chewed by the manatee's strong molars, which are its only teeth. When the manatee is born, each jaw has two vestigial incisors, which the manatee loses as it matures. If the African manatee's molars happen to fall out, new molars grow in their place. The manatee's flippers, which have nails, are also used to graze other manatees. The African manatee does not have any hind limbs. From the exterior, the African manatee looks very similar to the American manatee; however, the African manatee is different from the Amazonian manatee, which has characteristic white markings on its abdomen.

Taxonomy

The African manatee was officially declared a species under the Trichechus senegalensis taxon in 1795 by naturalist Johann Heinrich Friedrich Link. No subspecies of this taxon are known. Although African manatees live in both coastal areas and isolated inland areas, genetic evidence suggests no significant differences between the two populations. The African manatee falls under the genus Trichechus with only two other species, the Amazonian manatee and the West Indian manatee, which are also sirenians.

Range and Habitat

African manatees can be found in West African regions: Angola, Benin, Cameroon, Chad, the Republic of the Congo, the Democratic Republic of the Congo, Côte d'Ivoire, Equatorial Guinea, Gabon, The Gambia, Ghana, Guinea, Guinea-Bissau, Liberia, Mali, Mauritania, Niger, Nigeria, Senegal, Sierra Leone, and Togo. Manatees are found in brackish waters to freshwater: in oceans, rivers, lakes, coastal estuaries, reservoirs, lagoons, and bays on the coast. African manatees rarely inhabit waters with a temperature below 18 °C (64 °F).

Manatees have been found as far as 75 kilometres (47 mi) offshore, where there are shallow coastal flats and calm mangrove creeks filled with seagrass. Inland lakes where manatees dwell include Lake Volta, the Inner Niger River Delta in Mali, Lake Léré, and Lake de Tréné. Due to fluctuating flow rates and water levels in rivers, some of these permanent lakes serve as refuges for manatees in connecting rivers during the dry season. From north to south, the river systems that contain manatees include: the Senegal, Saloum, Gambia, Casamance, Cacheu, Mansôa, Geba, Buba, Tombali, Cacine, Kogon, Kondoure, Sierra Leone, Great Scarcies, Little Scarcies, Sherbro, Malem, Waanje, Sewa, Missunado, Cavalla, St. Paul, Morro, St. John, Bandama, Niouniourou, Sassandra, Comoé, Bia, Tano, Volta, Mono, Oueme, Niger, Mekrou, Benue, Cross, Katsena Ala, Bani, Akwayafe, Rio del Rey, Ngosso, Andokat, Mene, Munaya, Wouri, Sanaga, Faro, Chari, Bamaingui, Bahr-Kieta, Logoné, Mitémélé, Gabon, Ogoué, Lovanzi, Kouilou, Congo, Dande, Bengo, and Cuanza. Manatees move up these rivers until they are unable to proceed because of shallow waters or strong waterfalls.

The areas with the highest manatee populations are Guinea-Bissau, the lagoons of Côte d'Ivoire, the southern portions of the Niger River in Nigeria, the Sanaga River in Cameroon, the coastal lagoons in Gabon, and the lower parts of the Congo River. As part of a study completed in Côte d'Ivoire to assess where the majority of African manatees favor living, a sample of African manatees was radio-tagged and tracked. The tracking observed most of the sample in coastal lagoons, mangroves, and other herbaceous growths. They were also found in the grassy estuaries of big rivers with mangroves and in protected coastal spots with less than 3 metres (10 ft) of water containing both mangroves and marine macrophytes.

Diet

Manatees are herbivores; however, they also eat clams, mollusks, and fish found in nets. The percentage of the diet that is composed of non-plant material varies based on location. With manatees living off the coast having a lifetime average of 50% non-plant material. The West African manatee is the only Sirenian that seems to intentionally consume non-plant material. A majority of the African manatee's diet is made up of a variety of flora found above or hanging over the water. African manatees that inhabit rivers mostly eat the overhanging plants growing on the river banks. The diet of African manatees living in estuaries consists solely of mangrove trees. Each day, the African manatee eats about four to nine percent of its body weight in wet vegetation. Microorganisms within the African manatee's large intestine, which measures up to 20 metres or 66 feet in length, aid it in digesting the large quantity and variety of vegetation that it consumes daily.

Behavior

The African manatee is nocturnal. They tend to travel silently, eat, and be active towards the end of the day and during the nighttime. During the daytime, the African manatee dozes in shallow (1 to 2 meter deep) water. In countries such as Sierra Leone, African manatees migrate upstream when flooding occurs in June and July. This flooding can lower the availability of food for the manatees as well as lower the salinity of waterways. African manatees live in groups of 1 to 6. They have very few natural predators, two of which are sharks and crocodiles. They are also very social, spending a majority of their day bonding by touch, verbal communication, and smell. This creates a deep bond between them. When it is time to migrate due to a weather change, manatees will travel in larger groups to find warmer water and food.

Reproduction

The sex of an individual African manatee can only be determined by close examination of the manatee's underside. The only visible distinction between males and females is the genital openings. However, males tend to be smaller than females. Some female African manatees are sexually mature as young as 3 years of age, and they give birth every 3 to 5 years of their estimated 30-year lifespan. Males take a longer time to mature (about 9 to 10 years) and can rarely fertilize an egg at the age of 2 or 3 years. African manatees breed year-round. When males and females mate, it is not monogamous; multiple males will usually mate with one female. When the opportunity to mate with a female is at stake, males will fight with each other by pushing and shoving. Female African manatees give birth to one calf at a time after about a 13-month pregnancy. Calves can swim on their own at birth. Although the African manatee's social organization is not well understood, research shows the most common and tightly knit bonds are between a mother and her calf.

Threats

The African manatee is a vulnerable species because of its meat, oil, bones, and skin, which can bring great wealth to poachers. Specifically they are used to make walking sticks and toy spinning tops. In some countries, such as Nigeria and Cameroon, African manatees are sold to zoos, aquariums, online as pets, and they are sometimes shipped internationally. Anyone visiting such countries will notice manatee meat being sold on the streets and in marketplaces, but the lack of law enforcement protects the poachers from punishment. Residents of countries such as Mali and Chad depend on the oil of the African manatee to cure ailments such as ear infections, rheumatism, and skin conditions.

There are even more threats to the African manatees' habitat and life: urban and agricultural development, increased damming, and increased use of hydroelectric power in the rivers of countries like Côte d'Ivoire and Ghana. The building of dams has led to genetic isolation of some populations. There is little data to show if this has any negative long-term effects on the population as a whole. At several hydroelectric dams including the Kanji dam on the Niger River and the Akosombo dam on the Volta River manatees have been caught and killed in the turbines and intake valves. Thick congestion of boats in waterways may cause the manatees to have deadly run-ins with the vessels. However, even natural occurrences, such as droughts and tidal changes, can often strand manatees in unsuitable habitats. Some are killed accidentally by fishing trawls and in nets which are intended for catching sharks.

Some behaviors of African manatees provoke humans to hunt them. When manatees become tangled in fishing nets, they can damage them. People in countries such as Sierra Leone believe that killing the manatees to reduce the species size lowers the chances of the fishing nets requiring expensive repairs. In addition, African manatees can destroy rice crops by drifting into fields during the rainy season.

Many of the African manatees that venture up the Niger River starve to death. At certain times each year, the Niger River dries up due to the hot temperatures and lack of rain. Many manatees migrate there during the rainy season. When the water dries up the manatees are unable to get to other bodies of water.

Conservation

From November 2004 until December 2007, the West African Manatee Conservation Project completed Phase I. During this phase, residents of six African countries (Mauritania, Senegal, The Gambia, Guinea, Guinea-Bissau, and Sierra Leone) created a database of previously unknown information about the species (such as population, economic value, and habitat range) by conducting surveys in their countries. Other African countries also contributed reports that broadened the collective knowledge of the African manatee. Because of the work done during this phase, the general public, young children, and experienced scientists alike are receiving better information than ever before as to how to protect the African manatees. Phase I also allowed for up-close examination of the African manatee's way of life through field work.

Due to the large-scale success of Phase I, a Phase II is to be enacted by Wetlands International. During Phase II, the information collected in Phase I will be even more widely distributed around the areas in which the African manatee lives. Phase II will focus on furthering the existing research and adjusting legislation and education.

According to Appendix II of the Convention on International Trade in Endangered Species (CITES), the African manatee is endangered. CITES states that trade in any species on the list, including the African manatee, is to be carefully monitored and terminated. Laws exist to protect the African manatee in every country in which it lives, but these laws are not well enforced. Due to this mass lack of enforcement and minimal education, the African manatee population is being steadily depleted.

DUGONG

The dugong is a medium-sized marine mammal. It is one of four living species of the order Sirenia, which also includes three species of manatees. It is the only living representative of the once-diverse family Dugongidae; its closest modern relative, Steller's sea cow (Hydrodamalis gigas), was hunted to extinction in the 18th century. The dugong is the only strictly herbivorous marine mammal.

The dugong is the only sirenian in its range, which spans the waters of some 40 countries and territories throughout the Indo-West Pacific. The dugong is largely dependent on seagrass communities for subsistence and is thus restricted to the coastal habitats which support seagrass meadows, with the largest dugong concentrations typically occurring in wide, shallow, protected areas such as bays, mangrove channels, the waters of large inshore islands and inter-reefal waters. The northern waters of Australia between Shark Bay and Moreton Bay are believed to be the dugong's contemporary stronghold.

Like all modern sirenians, the dugong has a fusiform body with no dorsal fin or hind limbs. The forelimbs or flippers are paddle-like. The dugong is easily distinguished from the manatees by its fluked, dolphin-like tail, but also possesses a unique skull and teeth. Its snout is sharply down-turned, an adaptation for feeding in benthic seagrass communities. The molar teeth are simple and peg-like unlike the more elaborate molar dentition of manatees.

The dugong has been hunted for thousands of years for its meat and oil. Traditional hunting still has great cultural significance in several countries in its modern range, particularly northern Australia and the Pacific Islands. The dugong's current distribution is fragmented, and many populations are believed to be close to extinction. The IUCN lists the dugong as a species vulnerable to extinction, while the Convention on International Trade in Endangered Species limits or bans the trade of derived products. Despite being legally protected in many countries, the main causes of population decline remain anthropogenic and include fishing-related fatalities, habitat degradation and hunting. With its long lifespan of 70 years or more, and slow rate of reproduction, the dugong is especially vulnerable to extinction.

Taxonomy

Dugong dugon is the only extant species of the family Dugongidae, and one of only four extant species of the Sirenia order, the others forming the manatee family. It was first classified by Müller in 1776 as Trichechus dugon, a member of the manatee genus previously defined by Linnaeus. It was later assigned as the type species of Dugong by Lacépède and further classified within its own family by Gray and subfamily by Simpson.

Dugongs and other sirenians are not closely related to other marine mammals, being more related to elephants. Dugongs and elephants share a monophyletic group with hyraxes and the aardvark, one of the earliest offshoots of eutherians. The fossil record shows sirenians appearing in the Eocene, where they most likely lived in the Tethys Ocean. The two extant families of sirenians are thought to have diverged in the mid-Eocene, after which the dugongs and their closest relative, the Steller's sea cow, split off from a common ancestor in the Miocene. The Steller's sea cow became extinct in the 18th century. No fossils exist of other members of the Dugongidae.

Molecular studies have been made on dugong populations using mitochondrial DNA. The results have suggested that the population of Southeast Asia is distinct from the others. Australia has two distinct maternal lineages, one of which also contains the dugongs from Africa and Arabia. Limited genetic mixing has taken place between those in Southeast Asia and those in Australia, mostly around Timor. One of the lineages stretches all the way from Moreton Bay to Western Australia, while the other only stretches from Moreton Bay to the Northern Territory. There is not yet sufficient genetic data to make clear boundaries between distinct groups.

Anatomy and Morphology

The dugong's body is large with a cylindrical shape that tapers at both ends. It has thick, smooth skin that is a pale cream colour at birth, but darkens dorsally and laterally to brownish-to-dark-grey with age. The colour of a dugong can change due to the growth of algae on the skin. The body is sparsely covered in short hair, a common feature among sirenians which may allow for tactile interpretation of their environment. These hairs are most developed around the mouth, which has a large horseshoe-shaped upper lip forming a highly mobile muzzle. This muscular upper lip aids the dugong in foraging.

The dugong's tail flukes and flippers are similar to those of dolphins. These flukes are raised up and down in long strokes to move the animal forward, and can be twisted to turn. The forelimbs are paddle-like flippers which aid in turning and slowing. The dugong lacks nails on its flippers, which are only 15% of a dugong's body length. The tail has deep notches.

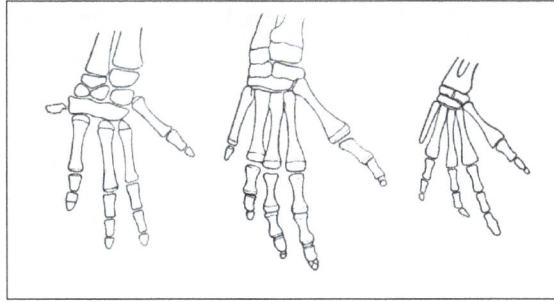
Bones in the forelimb can fuse variously with age.

A dugong's brain weighs a maximum of 300 g (11 oz), about 0.1% of the animal's body weight. With very small eyes, dugongs have limited vision, but acute hearing within narrow sound thresholds. Their ears, which lack pinnae, are located on the sides of their head. The nostrils are located on top of the head and can be closed using valves. Dugongs have two teats, one located behind each flipper. There are few differences between sexes; the body structures are almost the same. A male's testes are not externally located, and the main difference between males and females is the location of the genital aperture in relation to the umbilicus and the anus. The lungs in a dugong are very long, extending almost as far as the kidneys, which are also highly elongated in order to cope with the saltwater environment. If wounded, a dugong's blood will clot rapidly.

Dugong tail fluke.

The skull of a dugong is unique. The skull is enlarged with sharply down-turned premaxilla, which are stronger in males. The spine has between 57 and 60 vertebrae. Unlike in manatees, the dugong's teeth do not continually grow back via horizontal tooth replacement. The dugong has two incisors (tusks) which emerge in males during puberty. The female's tusks continue to grow without emerging during puberty, sometimes erupting later in life after reaching the base of the premaxilla. The number of growth layer groups in a tusk indicates the age of a dugong, and the cheek-teeth move forward with age. The full dental formula of dugongs is $\frac{2.0.3.3}{3.1.3.3}$, meaning they have two incisors, three premolars, and three molars on each side of their upper jaw, and three incisors, one canine, three premolars, and three molars on each side of their lower jaw. Like other sirenians, the dugong experiences pachyostosis, a condition in which the ribs and other long bones are unusually solid and contain little or no marrow. These heavy bones, which are among the densest in the animal kingdom, may act as a ballast to help keep sirenians suspended slightly below the water's surface.

An adult's length rarely exceeds 3 metres (9.8 ft). An individual this long is expected to weigh around 420 kilograms (926 lb). Weight in adults is typically more than 250 kilograms (551 lb) and less than 900 kilograms (1,984 lb). The largest individual recorded was 4.06 metres (13.32 ft) long and weighed 1,016 kilograms (2,240 lb), and was found off the Saurashtra coast of west India. Females tend to be larger than males.

Distribution and Habitat

Dugong on the sea floor at Marsa Alam, Egypt.

Dugongs are found in warm coastal waters from the western Pacific Ocean to the eastern coast of Africa, along an estimated 140,000 kilometres (86,992 mi) of coastline between 26° and 27° degrees to the north and south of the equator. Their historic range is believed to correspond to that of seagrasses from the Potamogetonaceae and Hydrocharitaceae families. The full size of the former range is unknown, although it is believed that the current populations represent the historical limits of the range, which is highly fractured. Today populations of dugongs are found in the waters of 37 countries and territories. Recorded numbers of dugongs are generally believed to be lower than actual numbers, due to a lack of accurate surveys. Despite this, the dugong population is thought to be shrinking, with a worldwide decline of 20 percent in the last 90 years. They have disappeared from the waters of Hong Kong, Mauritius, and Taiwan, as well as parts of Cambodia, Japan, the Philippines and Vietnam. Further disappearances are likely.

Dugongs are generally found in warm waters around the coast with large numbers concentrated in wide and shallow protected bays. The dugong is the only strictly-marine herbivorous mammal, as all species of manatee utilise fresh water to some degree. Nonetheless, they can tolerate the brackish waters found in coastal wetlands, and large numbers are also found in wide and shallow mangrove channels and around leeward sides of large inshore islands, where seagrass beds are common. They are usually located at a depth of around 10 m (33 ft), although in areas where the continental shelf remains shallow dugongs have been known to travel more than 10 kilometres (6 mi) from the shore, descending to as far as 37 metres (121 ft), where deepwater seagrasses such as Halophila spinulosa are found. Special habitats are used for different activities. It has been observed that shallow waters are used as sites for calving, minimising the risk of predation. Deep waters may provide a thermal refuge from cooler waters closer to the shore during winter.

East Africa and South Asia

In the late 1960s, herds of up to 500 dugongs were observed off the coast of East Africa and nearby islands. Current populations in this area are extremely small, numbering 50 and below, and it is thought likely they will become extinct. The eastern side of the Red Sea is home to large populations numbering in the hundreds, and similar populations are thought to exist on the western side. In the 1980s, it was estimated there could be as many as 4,000 dugongs in the Red Sea. Dugong populations in Madagascar are poorly studied, but due to widespread exploitation it is thought they may have severely declined, with few surviving individuals. In Mozambique, most of the remaining local populations are very small and the largest (about 120 individuals) occurs at Bazaruto Island, but they have become rare in historical habitats such as in Maputo Bay and on Inhaca Island. In Tanzania, observations have recently been increased around the Mafia Island Marine Park where a hunt was intended by fishermen but failed in 2009. In the Seychelles, dugongs had been regarded as extinct in 18th century until a small number was discovered around the Aldabra Atoll. This population may belong to a different group than that distributed among the inner isles. Dugongs once thrived among the Chagos Archipelago and Sea Cow Island was named after the species, although the species no longer occurs in the region.

A highly isolated breeding population exists in the Marine National Park, Gulf of Kutch, the only remaining population in western India. It is 1,500 kilometres (932 mi) from the population in the Persian Gulf, and 1,700 kilometres (1,056 mi) from the nearest population in India. Former populations in this area, centred on the Maldives and the Laccadive Islands, are presumed to be extinct. A population exists in the Gulf of Mannar Marine National Park and the Palk Strait between India and Sri Lanka, but it is seriously depleted. Recoveries of seagrass beds along former ranges of dugongs, such as the Chilika Lake have been confirmed in recent years, rising hopes for re-colorizations of the species. The population around the Andaman and Nicobar Islands are known only from a few records, and although the population was large during British rule, it is now believed to be small and scattered. Once distributed throughout the coastal belt in Sri Lanka, the dugong numbers have declined in last two decades.

Persian Gulf

The Persian Gulf has the second-largest dugong population in the world, inhabiting most of the southern coast, and the current population is believed to range from 5,800 to 7,300. In the course of a study being carried out in 1986 and 1999 on the Persian Gulf, the largest reported group sighting was made of more than 600 individuals to the west of Qatar. However, recent studies revealed severe declines both in population size and distributions among the region. A 2017 study, for instance, found a nearly 25% drop in population since 1950. Reasons for this drastic population loss include illegal poaching, oil spills and net entanglement.

Southern Pacific Outside of Australia

Dugong with attached remora off Lamen Island, Vanuatu.

A small population exists today along the southern coast of China, where efforts are being made to protect it, including the establishment of a seagrass sanctuary for dugong and other endangered marine fauna ranging in Guangxi. Despite these efforts, numbers continue to decrease, and in 2007 it was reported that no more dugong could be found on the west coast of the island of Hainan. Historically, dugongs were also present in the southern parts of the Yellow Sea.

In Vietnam, dugongs have been restricted mostly to the provinces of Kiên Giang and Bà Rịa–Vũng Tàu, including Phu Quoc Island and Con Dao Island, which hosted large populations in the past. Con Dao is now the only site in Vietnam where dugong are regularly seen, protected within the Côn Đảo National Park. Nonetheless, dangerously low levels of awarenesses for conservation of marine organisms in Vietnam and Cambodia may result in increased intentional or unintentional catches, and illegal trade is a potential danger for local dugongs. On Phu Quoc, the first 'Dugong Festival' was held in 2014, aiming to raise awareness of these issues.

In Thailand, the present distribution of dugongs is restricted to 6 provinces along the Andaman Sea, and very few dugongs are present in the Gulf of Thailand. The Gulf of Thailand was historically home to large number of the animals, but none have been sighted in the west of the gulf in recent years, and the remaining population in the east is thought to be very small and possibly declining. Dugongs are believed to exist in the Straits of Johor in very small numbers. The waters around Borneo support a small population, with more scattered throughout the Malay archipelago.

All the islands of the Philippines once provided habitats for sizeable herds of dugongs. They were common until the 1970s, when their numbers declined sharply due to accidental drownings in fishing gear and habitat destruction of seagrass meadows. Today, only isolated populations survive, most notably in the waters off the Calamian Islands in Palawan, Isabela in Luzon, Guimaras, and Mindanao. The dugong became the first marine animal protected by Philippine law, with harsh penalties for harming them.

Populations also exist around the Solomon Islands archipelago and New Caledonia, stretching to an easternmost population in Vanuatu. A highly isolated population lives around the islands of Palau.

A single dugong lives at Cocos (Keeling) Islands although the animal is thought to be a vagrant.

Northern Pacific

Today, possibly the smallest and northernmost population of dugongs exists around the Ryukyu islands, and a population formerly existed off Taiwan. An endangered population of 50 or fewer dugongs, possibly as few as three individuals, survives around Okinawa. New sightings of a cow-calf pair have been reported in 2017, indicating a possible breeding had occurred in these waters. A single individual was recorded at Amami Ōshima, at the northernmost edge of the dugong's historic range, more than 40 years after the last previous recorded sighting. A vagrant strayed into port near Ushibuka, Kumamoto, and died due to poor health. Historically, the Yaeyama Islands held a large concentration of dugongs, with more than 300 individuals. On Aragusuku Island, large quantities of skulls are preserved at an utaki that outsiders are strictly forbidden to enter. Dugong populations in these areas were reduced by historical hunts as payments to the Ryukyu Kingdom, before being wiped out because of large-scale illegal hunting and fishing using destructive methods such as dynamite fishing after the Second World War.

Populations around Taiwan appear to be almost extinct, although remnant individuals may visit areas with rich seagrass beds such as Dongsha Atoll. Some of the last reported sightings were made in Kenting National Park in 1950s and 60s. There had been occasional records of vagrants at the Northern Mariana Islands prior to 1985. It is unknown how much mixing there was between these populations historically. Some theorise that populations existed independently, for example that the Okinawan population were isolated members derived from the migration of a Philippine sub-species. Others postulate that the populations formed part of a super-population where migration between Ryukyu, Taiwan, and the Philippines was common.

Australia

Australia is home to the largest population, stretching from Shark Bay in Western Australia to Moreton Bay in Queensland. The population of Shark Bay is thought to be stable with over 10,000 dugongs. Smaller populations exist up the coast, including one in Ashmore reef. Large numbers of dugongs live to the north of the Northern Territory, with a population of over 20,000 in the gulf of Carpentaria alone. A population of over 25,000 exists in the Torres Strait such as off Thursday Island, although there is significant migration between the strait and the waters of New Guinea. The Great Barrier Reef provides important feeding areas for the species; this reef area houses a stable population of around 10,000, although the population concentration has shifted over time. Large bays facing north on the Queensland coast provide significant habitats for dugong, with the southernmost of these being Hervey Bay and Moreton Bay. Dugongs had been occasional visitors along the Gold Coast where an (re)establishment of a local population through range expansions has started recently.

Extinct Mediterranean Population

It has been confirmed that dugongs once inhabited the water of the Mediterranean possibly until after the rise of civilizations along the inland sea. This population possibly shared ancestry with the Red Sea population, and the Mediterranean population had never been large due to geographical factors and climate changes. The Mediterranean is the region where the Dugongidae originated in the mid-late Eocene, along with Caribbean Sea.

A mother and calf in shallow water.

Dugongs are long lived, and the oldest recorded specimen reached age 73. They have few natural predators, although animals such as crocodiles, killer whales, and sharks pose a threat to the young, and a dugong has also been recorded to have died from trauma after being impaled by a stingray barb. A large number of infections and parasitic diseases affect dugongs. Detected pathogens include helminths, cryptosporidium, different types of bacterial infections, and other unidentified parasites. 30% of dugong deaths in Queensland since 1996 are thought to be because of disease.

Although they are social animals, they are usually solitary or found in pairs due to the inability of seagrass beds to support large populations. Gatherings of hundreds of dugongs sometimes happen, but they last only for a short time. Because they are shy, and do not approach humans, little is known about dugong behaviour. They can go six minutes without breathing (though about two and a half minutes is more typical), and have been known to rest on their tail to breathe with their heads above water. They can dive to a maximum depth of 39 metres (128 ft); they spend most of their lives no deeper than 10 metres (33 ft). Communication between individuals is through chirps, whistles, barks, and other sounds that echo underwater. Different sounds have been observed with different amplitudes and frequencies, implying different purposes. Visual communication is limited due to poor eyesight, and is mainly used for activities such as lekking for courtship purposes. Mothers and calves are in almost constant physical contact, and calves have been known to reach out and touch their mothers with their flippers for reassurance.

Dugongs are semi-nomadic, often travelling long distances in search of food, but staying within a certain range their entire life. Large numbers often move together from one area to another. It is thought that these movements are caused by changes in seagrass availability. Their memory allows them to return to specific points after long travels. Dugong movements mostly occur within a localised area of seagrass beds, and animals in the same region show individualistic patterns of movement. Daily movement is affected by the tides. In areas where there is a large tidal range, dugongs travel with the tide in order to access shallower feeding areas. In Moreton Bay, dugongs often travel between foraging grounds inside the bay and warmer oceanic waters. At higher latitudes dugongs make seasonal travels to reach warmer water during the winter. Occasionally individual dugongs make long-distance travels over many days, and can travel over deep ocean waters. One animal was seen as far south as Sydney. Although they are marine creatures, dugongs have been known to travel up creeks, and in one case a dugong was caught 15 kilometres (9.3 mi) up a creek near Cooktown.

Feeding

Typical dugong feeding area in Moreton Bay.

Dugongs, along with other sirenians, are referred to as "sea cows" because their diet consists mainly of sea-grass. When eating they ingest the whole plant, including the roots, although when this is impossible they will feed on just the leaves. A wide variety of seagrass has been found in dugong stomach contents, and evidence exists they will eat algae when seagrass is scarce. Although almost completely herbivorous, they will occasionally eat invertebrates such as jellyfish, sea squirts, and shellfish. Dugongs in Moreton Bay, Australia, are omnivorous, feeding on invertebrates such as polychaetes or marine algae when the supply of their choice grasses decreases. In other southern areas of both western and eastern Australia, there is evidence that dugongs actively seek out large invertebrates. This does not apply to dugongs in tropical areas, in which faecal evidence indicates that invertebrates are not eaten.

Most dugongs do not feed from lush areas, but where the seagrass is more sparse. Additional factors such as protein concentration and regenerative ability also affect the value of a seagrass bed. The chemical structure and composition of the seagrass is important, and the grass species most often eaten are low in fibre, high in nitrogen, and easily digestible. In the Great Barrier Reef, dugongs feed on low-fibre high-nitrogen seagrass such as Halophila and Halodule, so as to maximize nutrient intake instead of bulk eating. Seagrasses of a lower seral are preferred, where the area has not fully vegetated. Only certain seagrass meadows are suitable for dugong consumption, due to the dugong's highly specialised diet. There is evidence that dugongs actively alter seagrass species compositions at local levels. Dugongs may search out deeper seagrass. Feeding trails have been observed as deep as 33 metres (108 ft), and dugongs have been seen feeding as deep as 37 metres (121 ft). Dugongs are relatively slow moving, swimming at around 10 kilometres per hour (6.2 mph). When moving along the seabed to feed they walk on their pectoral fins.

Dugong feeding may favor the subsequent growth low-fibre, high-nitrogen seagrasses such as Halophilia and Halodule. Species such as Zosteria capricorni are more dominant in established seagrass beds, but grow slowly, while Halophilia and Halodule grow quickly in the open space left by dugong feeding. This behavior is known as cultivation grazing, and favors the rapidly growing, higher nutrient seagrasses that dugongs prefer. Dugongs may also prefer to feed on younger, less fibrous strands of seagrasses, and cycles of cultivation feeding at different seagrass meadows may provide them with a greater number of younger plants.

Due to their poor eyesight, dugongs often use smell to locate edible plants. They also have a strong tactile sense, and feel their surroundings with their long sensitive bristles. They will dig up an entire plant and then shake it to remove the sand before eating it. They have been known to collect a pile of plants in one area before eating them. The flexible and muscular upper lip is used to dig out the plants. This leaves furrows in the sand in their path.

Reproduction and Parental Care

Dugong mother and offspring from East Timor.

A dugong reaches sexual maturity between the ages of eight and eighteen, older than in most other mammals. The way that females know how a male has reached sexual maturity is by the eruption of tusks in the male since tusks erupt in males when testosterone levels reach a high enough level. The age when a female first gives birth is disputed, with some studies placing the age between ten and seventeen years, while others place it as early as six years. There is evidence that male dugongs lose fertility at older ages. Despite the longevity of the dugong, which may live for 50 years or more, females give birth only a few times during their life, and invest considerable parental care in their young. The time between births is unclear, with estimates ranging from 2.4 to 7 years.

Mating behaviour varies between populations located in different areas. In some populations, males will establish a territory which females in heat will visit. In these areas a male will try to impress the females while defending the area from other males, a practice known as lekking. In other areas many males will attempt to mate with the same female, sometimes inflicting injuries to the female or each other. During this the female will have copulated with multiple males, who will have fought to mount her from below. This greatly increases the chances of conception.

Females give birth after a 13–15 month gestation, usually to just one calf. Birth occurs in very shallow water, with occasions known where the mothers were almost on the shore. As soon as the young is born the mother pushes it to the surface to take a breath. Newborns are already 1.2 metres (4 ft) long and weigh around 30 kilograms (66 lb). Once born, they stay close to their mothers, possibly to make swimming easier. The calf nurses for 14–18 months, although it begins to eat seagrasses soon after birth. A calf will only leave its mother once it has matured.

Importance to Humans

Dugongs have historically provided easy targets for hunters, who killed them for their meat, oil, skin, and bones. As the anthropologist A. Asbjørn Jøn has noted, they are often considered as the inspiration for mermaids, and people around the world developed cultures around dugong hunting. In some areas it remains an animal of great significance, and a growing ecotourism industry around dugongs has had economic benefit in some countries.

There is a 5,000-year-old wall painting of a dugong, apparently drawn by neolithic peoples, in Tambun Cave, Ipoh, Malaysia. This was discovered by Lieutenant R.L Rawlings in 1959 while on a routine patrol. During the Renaissance and the Baroque eras, dugongs were often exhibited in wunderkammers. They were also presented as Fiji mermaids in sideshows.

Dugong meat and oil have traditionally been some of the most valuable foods of Australian aborigines and Torres Strait Islanders. Some aborigines regard dugongs as part of their Aboriginality. Dugongs have also played a role in legends in Kenya, and the animal is known there as the "Queen of the Sea". Body parts are used as food, medicine, and decorations. In the Gulf states, dugongs served not only as a source of food, but their tusks were used as sword handles. Dugong oil is important as a preservative and conditioner for wooden boats to people in around the Gulf of Kutch in India, who also believe the meat to be an aphrodisiac. Dugong ribs were used to make carvings in Japan. In Southern China dugongs were traditionally regarded as a "miraculous fish", and it was bad luck to catch them. A wave of immigration beginning at the end of the 1950s resulted in dugongs being hunted for food. In the Philippines, dugongs are thought to bring bad luck, and parts of them are used to ward against evil spirits. In areas of Thailand it is believed that the dugong's tears form a powerful love potion, while in parts of Indonesia they are considered reincarnations of women. In Papua New Guinea they are seen as a symbol of strength.

Conservation

Dugong numbers have decreased in recent times. For a population to remain stable, 95 per cent of adults must survive the span of one year. The estimated percentage of females humans can kill without depleting the population is 1–2%. This number is reduced in areas where calving is minimal due to food shortages. Even in the best conditions a population is unlikely to increase more than 5% a year, leaving dugongs vulnerable to over-exploitation. The fact that they live in shallow waters puts them under great pressure from human activity. Research on dugongs and the effects of human activity on them has been limited, mostly taking place in Australia. In many countries, dugong numbers have never been surveyed. As such, trends are uncertain, with more data needed for comprehensive management. The only data stretching back far enough to mention population trends comes from the urban coast of Queensland, Australia. The last major worldwide study, made in 2002, concluded that the dugong was declining and possibly extinct in a third of its range, with unknown status in another half.

The IUCN Red List- lists the dugong as vulnerable, and the Convention on International Trade in Endangered Species of Wild Fauna and Flora regulates and in some areas has banned international trade. Regional cooperation is important due to the widespread distribution of the animal, and in 1998 there was strong support for Southeast Asian cooperation to protect dugongs. Kenya has passed legislation banning the hunting of dugongs and restricting trawling, but the dugong is not yet listed under Kenya's Wildlife Act for endangered species. Mozambique has had legislation to protect dugongs since 1955, but this has not been effectively enforced. Many marine parks have been established on the African coast of the Red Sea, and the Egyptian Gulf of Aqaba is fully protected. The United Arab Emirates has banned all hunting of dugongs within its waters, as has Bahrain. The UAE has additionally banned drift net fishing. India and Sri Lanka ban the hunting and selling of dugongs and their products. Japan has listed dugongs as endangered and has banned intentional kills and harassment. Hunting, catching, and harassment is banned by the People's Republic of China. The first marine mammal to be protected in the Philippines was the dugong, although monitoring this is difficult. Palau has legislated to protect dugongs, although this is not well enforced and poaching persists. Indonesia lists dugongs as a protected species, however protection is not always enforced and souvenir products made from dugong parts can be openly found in markets in Bali. The dugong is a national animal of Papua New Guinea, which bans all except traditional hunting. Vanuatu and New Caledonia ban hunting of dugongs. Dugongs are protected throughout Australia, although the rules vary by state; in some areas indigenous hunting is allowed. Dugongs are listed under the Nature Conservation Act in the Australian state of Queensland as vulnerable. Most currently live in established marine parks, where boats must travel at a restricted speed and mesh net fishing is restricted. In Vietnam, an illegal network targeting dugongs had been detected and was shut down in 2012. Potential hunts along Tanzanian coasts by fishermen have raised concerns as well.

Human Activity

Despite being legally protected in many countries, the main causes of population decline remain anthropogenic and include hunting, habitat degradation, and fishing-related fatalities. Entanglement in fishing nets has caused many deaths, although there are no precise statistics. Most issues with industrial fishing occur in deeper waters where dugong populations are low, with local fishing being the main risk in shallower waters. As dugongs cannot stay underwater for a very long period, they are highly prone to deaths due to entanglement. The use of shark nets has historically caused large numbers of deaths, and they have been eliminated in most areas and replaced with baited hooks. Hunting has historically been a problem too, although in most areas they are no longer hunted, with the exception of certain indigenous communities. In areas such as northern Australia, hunting remains the greatest impact on the dugong population.

Vessel strikes have proved a problem for manatees, but the relevance of this to dugongs is unknown. Increasing boat traffic has increased danger, especially in shallow waters. Ecotourism has increased in some countries, although effects remain undocumented. It has been seen to cause issues in areas such as Hainan due to environmental degradation. Modern farming practise and increased land clearing have also had an impact, and much of the coastline of dugong habitats is undergoing industrialisation, with increasing human populations. Dugongs accumulate heavy metal ions in their tissues throughout their lives, more so than other marine mammals. The effects are unknown. While international cooperation to form a conservative unit has been undertaken, socio-political needs are an impediment to dugong conservation in many developing countries. The shallow waters are often used as a source of food and income, problems exacerbated by aid used to improve fishing. In many countries, legislation does not exist to protect dugongs, and if it does it is not enforced.

Oil spills are a danger to dugongs in some areas, as is land reclamation. In Okinawa the small dugong population is threatened by United States military activity. Plans exist to build a military base close to the Henoko reef, and military activity also adds the threats of noise pollution, chemical pollution, soil erosion, and exposure to depleted uranium. The military base plans have been fought in US courts by some Okinawans, whose concerns include the impact on the local environment and dugong habitats. It was later revealed that the government of Japan was hiding evidence of the negative effects of ship lanes and human activities on dugongs observed during surveys carried out off Henoko reef. One of the three individuals has not been observed since June 2015, corresponding to the start of the excavation operations.

Environmental Degradation

If dugongs do not get enough to eat they may calve later and produce fewer young. Food shortages can be caused by many factors, such as a loss of habitat, death and decline in quality of seagrass, and a disturbance of feeding caused by human activity. Sewage, detergents, heavy metal, hypersaline water, herbicides, and other waste products all negatively affect seagrass meadows. Human activity such as mining, trawling, dredging, land reclamation, and boat propeller scarring also cause an increase in sedimentation which smothers seagrass and prevents light from reaching it. This is the most significant negative factor affecting seagrass.

Halophila ovalis—one of the dugong's preferred species of seagrass—declines rapidly due to lack of light, dying completely after 30 days. Extreme weather such as cyclones and floods can destroy hundreds of square kilometres of seagrass meadows, as well as washing dugongs ashore. The recovery of seagrass meadows and the spread of seagrass into new areas, or areas where it has been destroyed, can take over a decade. Most measures for protection involve restricting activities such as trawling in areas containing seagrass meadows, with little to no action on pollutants originating from land. In some areas water salinity is increased due to wastewater, and it is unknown how much salinity seagrass can withstand.

Dugong habitat in the Oura Bay area of Henoko, Okinawa, Japan, is currently under threat from land reclamation conducted by Japanese Government in order to build a US Marine base in the area. In August 2014, preliminary drilling surveys were conducted around the seagrass beds there. The construction is expected to seriously damage the dugong population's habitat, possibly leading to local extinction.

Capture and Captivity

The Australian state of Queensland has sixteen dugong protection parks, and some preservation zones have been established where even Aboriginal Peoples are not allowed to hunt. Capturing animals for research has caused only one or two deaths; dugongs are expensive to keep in captivity due to the long time mothers and calves spend together, and the inability to grow the seagrass that dugongs eat in an aquarium. Only one orphaned calf has ever been successfully kept in captivity.

Worldwide, only three dugongs are held in captivity. A female from the Philippines lives at Toba Aquarium in Toba, Mie, Japan. A male also lived there until he died on 10 February 2011. The second resides in Sea World Indonesia, after having been rescued from a fisherman's net and treated. The last one, a male, is kept at Sydney Aquarium, where he has resided since he was a juvenile. Sydney Aquarium had a second dugong for many years, until she died in 2018.

Gracie, a captive dugong at Underwater World, Singapore, was reported to have died in 2014 at the age of 19, from complications arising from an acute digestive disorder.

References

- Luiselli, L.; Akani, G.C.; Ebere, N.; Angelici, F. M.; Amori, G.; Politano, E. (2012). "Macro-habitat preferences by the African manatee and crocodiles – ecological and conservation implications". Web Ecology. 12 (1). Doi:10.5194/we-12-39-2012
- Sirenia, accounts: animaldiversity.org, Retrieved 12 May, 2019
- Shoshani, J. (2005). "Order Sirenia". In Wilson, D.E.; Reeder, D.M (eds.). Mammal Species of the World: A Taxonomic and Geographic Reference (3rd ed.). Johns Hopkins University Press. P. 93. ISBN 978-0-8018-8221-0. OCLC 62265494
- Ningthoujam Sandhyarani. "Interesting Facts about Manatees (Sea Cows)". Buzzle.com. Archived from the original on 2011-06-26. Retrieved 2011-06-28
- Deutsch, C.J.; Self-Sullivan, C. & Mignucci-Giannoni, A. (2008). "Trichechus manatus". The IUCN Red List of Threatened Species. 2008: e.T22103A9356917. Doi:10.2305/IUCN.UK.2008.RLTS.T22103A9356917.en
- "Manatee". Sea World. December 30, 2011. Archived from the original on January 18, 2012. Retrieved December 30, 2011

6

Otters and Polar Bears

Otters are the carnivorous mammals that are semiaquatic, aquatic or marine. Polar bears are hyper carnivorous bears that generally inhabit the Arctic Circle. They have a well-developed sense of smell as well as good vision. The topics elaborated in this chapter will help in gaining a better perspective about the various characteristics of otters and polar bears.

OTTERS

Otters, (subfamily Lutrinae), are any of 13 or 14 species of semiaquatic mammals that belong to the weasel family (Mustelidae) and are noted for their playful behaviour. The otter has a lithe and slender body with short legs, a strong neck, and a long flattened tail that helps propel the animal gracefully through water. Swimming ability is further enhanced in most species by four webbed feet. Two species are marine, with the others living predominantly in fresh water. Otters range in size from 3 kg (6.6 pounds) in the Asian small-clawed otter (Aonyx cinereus, formerly Amblonyx cinereus) to 26 kg (57 pounds) in the giant otter (Pteronura brasiliensis) and 45 kg (99 pounds) in the sea otter (Enhydra lutris). Fur colour is various shades of brown with lighter underparts.

Freshwater Otters

The 11 species often referred to as river otters are found throughout North America, South America, Europe, Africa, and Asia in freshwater ecosystems that sustain an abundance of prey such as fish, crayfish, crabs, mussels, and frogs. Most river otters are opportunistic, feeding on whatever is most easily obtained. Diet often varies seasonally or locally, depending on which prey is available. River otters hunt visually while chasing fish, but they use their manual dexterity to dislodge crabs and crayfish from under rocks. Sensory hairs on the snout called vibrissae also assist by sensing water turbulence. After being captured in the teeth or forefeet, prey is consumed either in the water or on shore. River otters hunt more effectively in shallow water than in deep water, and, although they are proficient swimmers, all prefer slow-swimming species of fish. African clawless otters (Aonyx capensis) and Congo clawless otters (A. congicus or A. capensis congicus) occupy murky waterways and thus rely more on manual dexterity than on vision to obtain food (mostly crabs) from under rocks. Their front feet are handlike and partially webbed.

Most travel is aquatic, but river otters can venture swiftly overland between bodies of water. They typically follow the shortest route possible and often establish much-used trails. While in the water, they constantly search features such as logjams and deepwater pools for prey. To rest, otters seek refuge in underground holes, rock crevices, beaver lodges, cavities in root systems, or simply dense vegetation along the shoreline. When not resting or eating, river otters can often be seen eagerly sliding down mud or snow banks. Many species establish regular latrine sites along the shores of lakes or rivers. Such stations may facilitate communication between individuals.

North American river otters (Lutra canadensis)
explore a melting river and catch fish.

Litter size ranges from one to five. Young otters (pups) may fall prey to large raptors, and various carnivores may kill adults traveling on land. In warmer regions crocodiles and alligators are threats. However, most mortality results from human activities, in the form of road kills, drownings in fishnets, destruction as pests around fishing areas, or trapping for their fur.

Saltwater Otters

Two otter species are strictly marine: the sea otter (Enhydra lutris) of the Pacific Coast of North America and the much smaller marine otter (Lontra felina) from the coast of Peru and Chile. Both rely exclusively on marine prey, although the sea otter can be found much farther offshore; the marine otter stays within about 100 metres (330 feet) of the shore.

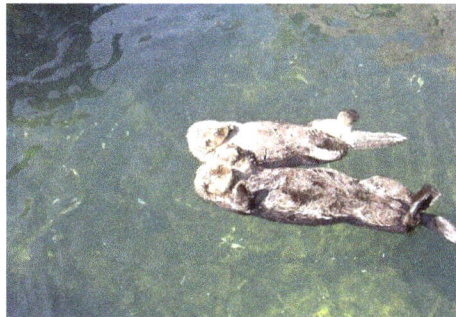

Sea otters at the Vancouver Aquarium, Stanley Park,
Vancouver, British Columbia, Canada.

Sea otters are well adapted to marine life. The front and back feet are fully webbed; large lungs allow long dives and provide buoyancy and thick fur provides insulation. Sea otters can also drink salt waterand thus can remain at sea for several days at a time. Sea otters are usually solitary but are sometimes seen in groups—gatherings of up to 2,000 have been observed along the coast of Alaska. At night, sea otters may choose either to sleep on land or simply to rest afloat near beds of kelp. They feed mainly on sea urchins, crabs, and various shellfish. Fish are also eaten. Captured prey is eaten at sea while the otter swims on its back. Rocks are typically used to break open crabs and shellfish, whereas sea urchins are crushed with the forefeet and teeth. Sea otter predation on the herbivorous urchins (genus Strongylocentrotus) enables kelp forests and the fish associated with them to flourish. However, large numbers of sea otters can deplete shellfish populations, conflicting with fisheries for crabs, clams, and abalones. Females give birth in water to only one young, which remains dependent on the mother until six to eight months of age. Sharks and killer whales eat sea otters on occasion.

The marine otter is really a freshwater otter that has learned to occupy marine environments in South America. This small (3–6 kg [6.6–13.2 pounds]) otter occurs on the Pacific coast from Peru through Chile and Tierra del Fuego in Argentina. It is mostly solitary, and only rarely do groups of more than three animals occur. Marine otters occupy the intertidal zone that covers the first 100–150 metres (roughly 330–500 feet) of coastal water and about 30 metres (100 feet) inland. They feed on crustaceans such as crabs and shrimp, as well as mollusks and fish. Interestingly, marine otters do not consume sea urchins as extensively as do sea otters, even though sea urchins often are common where marine otters live. Unlike sea otters, marine otters shelter in rock cavities for daytime rest and parturition.

Conservation and Classification

Nearly all species of otters face increasing threat as urbanization and logging continue. North American river otters (L. canadensis) are still taken as part of the commercial fur trade, but the primary threats to others are the destruction of wetland habitats and pollution. Heavy metals and contaminants such as mercury and PCBs accumulate in otter tissues and in time impair both reproduction and survival. Pollution also affects fish populations on which otters often depend. Conservation of remaining wetlands and restoration of water quality are currently the most important steps toward ensuring the future of otters.

Most authorities maintain that 13 species of otters make up the subfamily Lutrinae. The status of the Congo clawless otter remains a subject of debate, however, with most researchers considering the animal to be a subspecies of the African small-clawed otter (Aonyx capensis) and giving it the taxonomic name A. capensis congicus. Others claim that the Congo clawless otter is a valid species and have given it the taxonomic name A. congicus.

POLAR BEARS

The polar bear (*Ursus maritimus*) is a hypercarnivorous bear whose native range lies largely within the Arctic Circle, encompassing the Arctic Ocean, its surrounding seas and surrounding land masses. It is a large bear, approximately the same size as the omnivorous Kodiak bear (*Ursus arctos middendorffi*). A boar (adult male) weighs around 350–700 kg (772–1,543 lb), while a sow (adult female) is about half that size. Polar bears are the largest land carnivores currently in existence, rivaled only by the Kodiak bear.Although it is the sister species of the brown bear, it has evolved to occupy a narrower ecological niche, with many body characteristics adapted for cold temperatures, for moving across snow, ice and open water, and for hunting seals, which make up most of its diet. Although most polar bears are born on land, they spend most of their time on the sea ice. Their scientific name means "maritime bear" and derives from this fact. Polar bears hunt their preferred food of seals from the edge of sea ice, often living off fat reserves when no sea ice is present. Because of their dependence on the sea ice, polar bears are classified as marine mammals.

Because of expected habitat loss caused by climate change, the polar bear is classified as a vulnerable species. For decades, large-scale hunting raised international concern for the future of the species, but populations rebounded after controls and quotas began to take effect. For thousands of years, the polar bear has been a key figure in the material, spiritual, and cultural life of circumpolar peoples, and polar bears remain important in their cultures. Historically, the polar bear has also been known as the white bear. It is sometimes referred to as the nanook, based on the Inuit term nanuq.

Constantine John Phipps was the first to describe the polar bear as a distinct species in 1774. He chose the scientific name Ursus maritimus, the Latin for 'maritime bear', due to the animal's native habitat. The Inuit refer to the animal as nanook. The Yupik also refer to the bear as nanuuk in Siberian Yupik. The bear is umka in the Chukchi language. In the Norwegian-administered Svalbard archipelago, the polar bear is referred to as Isbjørn ("ice bear").

The polar bear was previously considered to be in its own genus, Thalarctos. However, evidence of hybrids between polar bears and brown bears, and of the recent evolutionary divergence of the two species, does not support the establishment of this separate genus, and the accepted scientific name is now therefore Ursus maritimus, as Phipps originally proposed.

Polar bears have evolved adaptations for Arctic life. For example,
large furry feet and short, sharp, stocky claws give them good traction on ice.

The bear family, Ursidae, is thought to have split from other carnivorans about 38 million years ago. The subfamily Ursinae originated approximately 4.2 million years ago. The oldest known polar bear fossil is a 130,000 to 110,000-year-old jaw bone, found on Prince Charles Foreland in 2004. Fossils show that between 10,000 and 20,000 years ago, the polar bear's molar teeth changed significantly from those of the brown bear. Polar bears are thought to have diverged from a population of brown bears that became isolated during a period of glaciation in the Pleistocene from the eastern part of Siberia (from Kamchatka and the Kolym Peninsula).

The evidence from DNA analysis is more complex. The mitochondrial DNA (mtDNA) of the polar bear diverged from the brown bear, *Ursus arctos*, roughly 150,000 years ago. Further, some clades of brown bear, as assessed by their mtDNA, are more closely related to polar bears than to other brown bears, meaning that the polar bear might not be considered a species under some species concepts. The mtDNA of extinct Irish brown bears is particularly close to polar bears. A comparison of the nuclear genome of polar bears with that of brown bears revealed a different pattern, the two forming genetically distinct clades that diverged approximately 603,000 years ago, although the latest research is based on analysis of the complete genomes (rather than just the mitochondria or partial nuclear genomes) of polar and brown bears, and establishes the divergence of polar and brown bears at 400,000 years ago.

However, the two species have mated intermittently for all that time, most likely coming into contact with each other during warming periods, when polar bears were driven onto land and brown bears migrated northward. Most brown bears have about 2 percent genetic material from polar bears, but one population, the ABC Islands bears has between 5 percent and 10 percent polar bear genes, indicating more frequent and recent mating. Polar bears can breed with brown bears to produce fertile grizzly–polar bear hybrids; rather than indicating that they have only recently diverged, the new evidence suggests more frequent mating has continued over a longer period of time, and thus the two bears remain genetically similar. However, because neither species can survive long in the other's ecological niche, and because they have different morphology, metabolism, social and feeding behaviours, and other phenotypic characteristics, the two bears are generally classified as separate species.

When the polar bear was originally documented, two subspecies were identified: the American polar bear (*Ursus maritimus maritimus*) by Constantine J. Phipps in 1774, and the Siberian polar bear (*Ursus maritimus marinus*) by Peter Simon Pallas in 1776. This distinction has since been invalidated. One alleged fossil subspecies has been identified: *Ursus maritimus tyrannus*, which became extinct during the Pleistocene. *U.m. tyrannus* was significantly larger than the living subspecies. However, recent reanalysis of the fossil suggests that it was actually a brown bear.

Population and Distribution

Polar bears investigate the submarine USS Honolulu
450 kilometres (280 mi) from the North Pole.

The polar bear is found in the Arctic Circle and adjacent land masses as far south as Newfoundland. Due to the absence of human development in its remote habitat, it retains more of its original range than any other extant carnivore. While they are rare north of 88°, there is evidence that they range all the way across the Arctic, and as far south as James Bay in Canada. Their southernmost range is near the boundary between the subarctic and humid continental climate zones. They can occasionally drift widely with the sea ice, and there have been anecdotal sightings as far south as Berlevåg on the Norwegian mainland and the Kuril Islands in the Sea of Okhotsk. It is difficult to estimate a global population of polar bears as much of the range has been poorly studied; however, biologists use a working estimate of about 20–25,000 or 22–31,000 polar bears worldwide.

There are 19 generally recognized, discrete subpopulations, though polar bears are thought to exist only in low densities in the area of the Arctic Basin. The subpopulations display seasonal fidelity to particular areas, but DNA studies show that they are not reproductively isolated. The 13 North American subpopulations range from the Beaufort Sea south to Hudson Bay and east to Baffin Bay in western Greenland and account for about 54% of the global population.

The range includes the territory of five nations: Denmark (Greenland), Norway (Svalbard), Russia, the United States (Alaska) and Canada. These five nations are the signatories of the International Agreement on the Conservation of Polar Bears, which mandates cooperation on research and conservation efforts throughout the polar bear's range. Bears sometimes swim to Iceland from Greenland—about 600 sightings since the country's settlement in the 9th century AD, and five in the 21st century as of 2016—and are always killed because of their danger, and the cost and difficulty of repatriation.

Modern methods of tracking polar bear populations have been implemented only since the mid-1980s, and are expensive to perform consistently over a large area. The most accurate counts require flying a helicopter in the Arctic climate to find polar bears, shooting a tranquilizer dart at the bear to sedate it, and then tagging the bear. In Nunavut, some Inuit have reported increases in bear sightings around human settlements in recent years, leading to a belief that populations are increasing. Scientists have responded by noting that hungry bears may be congregating around human settlements, leading to the illusion that populations are higher than they actually are. The Polar Bear Specialist Group of the IUCN Species Survival Commission takes the position that "estimates of subpopulation size or sustainable harvest levels should not be made solely on the basis of traditional ecological knowledge without supporting scientific studies."

Of the 19 recognized polar bear subpopulations, one is in decline, two are increasing, seven are stable, and nine have insufficient data, as of 2017.

Habitat

Polar bear jumping on fast ice.

The polar bear is a marine mammal because it spends many months of the year at sea. However, it is the only living marine mammal with powerful, large limbs and feet that allow them to cover miles on foot and run on land. Its preferred habitat is the annual sea ice covering the waters over the continental shelf and the Arctic inter-island archipelagos. These areas, known as the "Arctic ring of life", have high biological productivity in comparison to the deep waters of the high Arctic. The polar bear tends to frequent areas where sea ice meets water, such as polynyas and leads (temporary stretches of open water in Arctic ice), to hunt the seals that make up most of its diet. Freshwater is limited in these environments because it is either locked up in snow or saline. Polar bears are able to produce water through the metabolism of fats found in seal blubber. Polar bears are therefore found primarily along the perimeter of the polar ice pack, rather than in the Polar Basin close to the North Pole where the density of seals is low.

Annual ice contains areas of water that appear and disappear throughout the year as the weather changes. Seals migrate in response to these changes, and polar bears must follow their prey. In Hudson Bay, James Bay, and some other areas, the ice melts completely each summer (an event often referred to as "ice-floe breakup"), forcing polar bears to go onto land and wait through the months until the next freeze-up. In the Chukchi and Beaufort seas, polar bears retreat each summer to the ice further north that remains frozen year-round.

Physical Characteristics

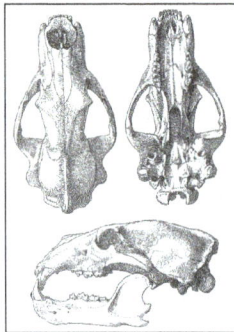

Skull, as illustrated by N. N. Kondakov.

Skull of a polar bear.

The only other bear similar in size to the polar bear is the Kodiak bear, which is a subspecies of brown bear. Adult male polar bears weigh 350–700 kg (772–1,543 lb) and measure 2.4–3 metres (7 ft 10 in–9 ft 10 in) in total length. Around the Beaufort Sea, however, mature males reportedly average 450 kg (992 lb). Adult females are roughly half the size of males and normally weigh 150–250 kg (331–551 lb), measuring 1.8–2.4 metres (5 ft 11 in–7 ft 10 in) in length. Elsewhere, a slightly larger estimated average weight of 260 kg (573 lb) was claimed for adult females. When pregnant, however, females can weigh as much as 500 kg (1,102 lb). The polar bear is among the most sexually dimorphic of mammals, surpassed only by the pinnipeds such as elephant seals. The largest polar bear on record, reportedly weighing 1,002 kg (2,209 lb), was a male shot at Kotzebue Sound in northwestern Alaska in 1960. This specimen, when mounted, stood 3.39 m (11 ft 1 in) tall on its hindlegs. The shoulder height of an adult polar bear is 122 to 160 cm (4 ft 0 in to 5 ft 3 in). While all bears are short-tailed, the polar bear's tail is relatively the shortest amongst living bears, ranging from 7 to 13 cm (2.8 to 5.1 in) in length.

Compared with its closest relative, the brown bear, the polar bear has a more elongated body build and a longer skull and nose. As predicted by Allen's rule for a northerly animal, the legs are stocky and the ears and tail are small. However, the feet are very large to distribute load when walking on snow or thin ice and to provide propulsion when swimming; they may measure 30 cm (12 in) across in an adult. The pads of the paws are covered with small, soft papillae (dermal bumps), which provide traction on the ice. The polar bear's claws are short and stocky compared to those of the brown bear, perhaps to serve the former's need to grip heavy prey and ice. The claws are deeply scooped on the underside to assist in digging in the ice of the natural habitat. Research of injury patterns in polar bear forelimbs found injuries to the right forelimb to be more frequent than those to the left, suggesting, perhaps, right-handedness. Unlike the brown bear, polar bears in captivity are rarely overweight or particularly large, possibly as a reaction to the warm conditions of most zoos.

The 42 teeth of a polar bear reflect its highly carnivorous diet. The cheek teeth are smaller and more jagged than in the brown bear, and the canines are larger and sharper. The dental formula is $\frac{3.1.42}{3.1.43}$.

Polar bears are superbly insulated by up to 10 cm (4 in) of adipose tissue, their hide and their fur; they overheat at temperatures above 10 °C (50 °F), and are nearly invisible under infrared photography. Polar bear fur consists of a layer of dense underfur and an outer layer of guard hairs, which appear white to tan but are actually transparent. Two genes that are known to influence melanin production, LYST and AIM1, are both mutated in polar bears, possibly leading to the absence on this pigment in their fur. The guard hair is 5–15 cm (2–6 in) over most of the body. Polar bears gradually moult from May to August, but, unlike other Arctic mammals, they do not shed their coat for a darker shade to provide camouflage in summer conditions. The hollow guard hairs of a polar bear coat were once thought to act as fiber-optic tubes to conduct light to its black skin, where it could be absorbed; however, this hypothesis was disproved by a study in 1998.

The white coat usually yellows with age. When kept in captivity in warm, humid conditions, the fur may turn a pale shade of green due to algae growing inside the guard hairs. Males have significantly longer hairs on their forelegs, which increase in length until the bear reaches 14 years of age. The male's ornamental foreleg hair is thought to attract females, serving a similar function to the lion's mane.

The polar bear has an extremely well developed sense of smell, being able to detect seals nearly 1.6 km (1 mi) away and buried under 1 m (3 ft) of snow. Its hearing is about as acute as that of a human, and its vision is also good at long distances.

The polar bear is an excellent swimmer and often will swim for days. One bear swam continuously for 9 days in the frigid Bering Sea for 700 km (400 mi) to reach ice far from land. She then travelled another 1,800 km (1,100 mi). During the swim, the bear lost 22% of her body mass and her yearling cub died. With its body fat providing buoyancy, the bear swims in a dog paddle fashion using its large forepaws for propulsion. Polar bears can swim 10 km/h (6 mph). When walking, the polar bear tends to have a lumbering gait and maintains an average speed of around 5.6 km/h (3.5 mph). When sprinting, they can reach up to 40 km/h (25 mph).

Life and Behaviour

Unlike brown bears, polar bears are not territorial. Although stereotyped as being voraciously aggressive, they are normally cautious in confrontations, and often choose to escape rather than fight. Satiated polar bears rarely attack humans unless severely provoked. However, due to their lack of prior human interaction, hungry polar bears are extremely unpredictable, fearless towards people and are known to kill and sometimes eat humans. Many attacks by brown bears are the result of surprising the animal, which is not the case with the polar bear. Polar bears are stealth hunters, and the victim is often unaware of the bear's presence until the attack is underway. Whereas brown bears often maul a person and then leave, polar bear attacks are more likely to be predatory and are almost always fatal. However, due to the very small human population around the Arctic, such attacks are rare. Michio Hoshino, a Japanese wildlife photographer, was once pursued briefly by a hungry male polar bear in northern Alaska. According to Hoshino, the bear started running but Hoshino made it to his truck. The bear was able to reach the truck and tore one of the doors off the truck before Hoshino was able to drive off.

Subadult polar bear males frequently play-fight. During the mating season,
actual fighting is intense and often leaves scars or broken teeth.

In general, adult polar bears live solitary lives. Yet, they have often been seen playing together for hours at a time and even sleeping in an embrace, and polar bear zoologist Nikita Ovsianikov has described adult males as having "well-developed friendships." Cubs are especially playful as well. Among young males in particular, play-fighting may be a means of practicing for serious competition during mating seasons later in life. Polar bears are usually quiet but do communicate with various sounds and vocalizations. Females communicate with their young with moans and chuffs, and the distress calls of both cubs and subadults consists of bleats. Cubs may hum while nursing. When nervous, bears produce huffs, chuffs and snorts while hisses, growls and roars are signs of aggression. Chemical communication can also be important: bears leave behind their scent in their tracks which allow individuals to keep track of one another in the vast Arctic wilderness.

In 1992, a photographer near Churchill took a now widely circulated set of photographs of a polar bear playing with a Canadian Eskimo Dog (*Canis lupus familiaris*) a tenth of its size. The pair wrestled harmlessly together each afternoon for 10 days in a row for no apparent reason, although the bear may have been trying to demonstrate its friendliness in the hope of sharing the kennel's food. This kind of social interaction is uncommon; it is far more typical for polar bears to behave aggressively towards dogs.

Hunting and Diet

The polar bear is the most carnivorous member of the bear family, and throughout most of its range, its diet primarily consists of ringed (*Pusa hispida*) and bearded seals (*Erignathus barbatus*). The

Arctic is home to millions of seals, which become prey when they surface in holes in the ice in order to breathe, or when they haul out on the ice to rest. Polar bears hunt primarily at the interface between ice, water, and air; they only rarely catch seals on land or in open water.

Long muzzle and neck of the polar bear help it to search in deep holes
for seals, while powerful hindquarters enable it to drag massive prey.

The polar bear's most common hunting method is called *still-hunting*: the bear uses its excellent sense of smell to locate a seal breathing hole, and crouches nearby in silence for a seal to appear. The bear may lay in wait for several hours. When the seal exhales, the bear smells its breath, reaches into the hole with a forepaw, and drags it out onto the ice. The polar bear kills the seal by biting its head to crush its skull. The polar bear also hunts by stalking seals resting on the ice: upon spotting a seal, it walks to within 90 m (100 yd), and then crouches. If the seal does not notice, the bear creeps to within 9 to 12 m (30 to 40 ft) of the seal and then suddenly rushes forth to attack. A third hunting method is to raid the birth lairs that female seals create in the snow.

A widespread legend tells that polar bears cover their black noses with their paws when hunting. This behaviour, if it happens, is rare – although the story exists in the oral history of northern peoples and in accounts by early Arctic explorers, there is no record of an eyewitness account of the behaviour in recent decades.

Polar bear feeding on a bearded seal.

Mature bears tend to eat only the calorie-rich skin and blubber of the seal, which are highly digestible, whereas younger bears consume the protein-rich red meat. Studies have also photographed polar bears scaling near-vertical cliffs, to eat birds' chicks and eggs. For subadult bears, which are independent of their mother but have not yet gained enough experience and body size to successfully hunt seals, scavenging the carcasses from other bears' kills is an important source of nutrition. Subadults may also be forced to accept a half-eaten carcass if they kill a seal but cannot defend it from larger polar bears. After feeding, polar bears wash themselves with water or snow.

Although polar bears are extraordinarily powerful, its primary prey species, the ringed seal, is much smaller than itself, and many of the seals hunted are pups rather than adults. Ringed seals are born weighing 5.4 kg (12 lb) and grown to an estimated average weight of only 60 kg (130 lb). They also in places prey heavily upon the harp seal (*Pagophilus groenlandicus*) or the harbor seal. The bearded seal, on the other hand, can be nearly the same size as the bear itself, averaging 270 kg (600 lb). Adult male bearded seals, at 350 to 500 kg (770 to 1,100 lb) are too large for a female bear to overtake, and so are potential prey only for mature male bears. Large males also occasionally attempt to hunt and kill even larger prey items. It can kill an adult walrus (*Odobenus rosmarus*), although this is rarely attempted. At up to 2,000 kg (4,400 lb) and a typical adult mass range of 600 to 1,500 kg (1,300 to 3,300 lb), a walrus can be more than twice the bear's weight, has extremely thick skin and has up to 1-metre (3 ft)-long ivory tusks that can be used as formidable weapons. A polar bear may charge a group of walruses, with the goal of separating a young, infirm, or injured walrus from the pod. They will even attack adult walruses when their diving holes have frozen over or intercept them before they can get back to the diving hole in the ice. Yet, polar bears will very seldom attack full-grown adult walruses, with the largest male walrus probably invulnerable unless otherwise injured or incapacitated. Since an attack on a walrus tends to be an extremely protracted and exhausting venture, bears have been known to back down from the attack after making the initial injury to the walrus. Polar bears have also been seen to prey on beluga whales (*Delphinapterus leucas*) and narwhals (*Monodon monoceros*), by swiping at them at breathing holes. The whales are of similar size to the walrus and nearly as difficult for the bear to subdue. Most terrestrial animals in the Arctic can outrun the polar bear on land as polar bears overheat quickly, and most marine animals the bear encounters can outswim it. In some areas, the polar bear's diet is supplemented by walrus calves and by the carcasses of dead adult walruses or whales, whose blubber is readily devoured even when rotten. Polar bears sometimes swim underwater to catch fish like the Arctic charr or the fourhorn sculpin.

Some characteristic postures:

- At rest.
- Assessing a situation.
- When feeding.

With the exception of pregnant females, polar bears are active year-round, although they have a vestigial hibernation induction trigger in their blood. Unlike brown and black bears, polar bears are capable of fasting for up to several months during late summer and early fall, when they cannot hunt for seals because the sea is unfrozen. When sea ice is unavailable during summer and early autumn, some populations live off fat reserves for months at a time, as polar bears do not 'hibernate' any time of the year.

Being both curious animals and scavengers, polar bears investigate and consume garbage where they come into contact with humans. Polar bears may attempt to consume almost anything they can find, including hazardous substances such as styrofoam, plastic, car batteries, ethylene glycol, hydraulic fluid, and motor oil. The dump in Churchill, Manitoba was closed in 2006 to protect bears, and waste is now recycled or transported to Thompson, Manitoba.

Dietary Flexibility

Although seal predation is the primary and an indispensable way of life for most polar bears, when alternatives are present they are quite flexible. Polar bears consume a wide variety of other wild foods, including muskox (*Ovibos moschatus*), reindeer (*Rangifer tarandus*), birds, eggs, rodents, crabs, other crustaceans and other polar bears. They may also eat plants, including berries, roots, and kelp; however, none of these have been a significant part of their diet, except for beachcast marine mammal carcasses. Given the change in climate, with ice breaking up in areas such as the Hudson Bay earlier than it used to, polar bears are exploiting food resources such as snow geese and eggs, and plants such as lyme grass in increased quantities.

When stalking land animals, such as muskox, reindeer, and even willow ptarmigan (*Lagopus lagopus*), polar bears appear to make use of vegetative cover and wind direction to bring them as close to their prey as possible before attacking. Polar bears have been observed to hunt the small Svalbard reindeer (*R. t. platyrhynchus*), which weigh only 40 to 60 kg (90 to 130 lb) as adults, as well as the barren-ground caribou (*R. t. groenlandicus*), which is about twice as heavy as that. Adult muskox, which can weigh 450 kg (1,000 lb) or more, are a more formidable quarry. Although ungulates are not typical prey, the killing of one during the summer months can greatly increase the odds of survival during that lean period. Like the brown bear, most ungulate prey of polar bears is likely to be young, sickly or injured specimens rather than healthy adults. The polar bear's metabolism is specialized to require large amounts of fat from marine mammals, and it cannot derive sufficient caloric intake from terrestrial food.

In their southern range, especially near Hudson Bay and James Bay, Canadian polar bears endure all summer without sea ice to hunt from. Here, their food ecology shows their dietary flexibility. They still manage to consume some seals, but they are food-deprived in summer as only marine mammal carcasses are an important alternative without sea ice, especially carcasses of the beluga whale. These alternatives may reduce the rate of weight loss of bears when on land. One scientist found that 71% of the Hudson Bay bears had fed on seaweed (marine algae) and that about half were feeding on birds such as the dovekie and sea ducks, especially the long-tailed duck (53%) and common eider, by swimming underwater to catch them. They were also diving to feed on blue mussels and other underwater food sources like the green sea urchin. 24% had eaten moss recently, 19% had consumed grass, 34% had eaten black crowberry and about half had consumed willows. This study illustrates the polar bear's dietary flexibility but it does not represent its life history elsewhere. Most polar bears elsewhere will never have access to these alternatives, except for the marine mammal carcasses that are important wherever they occur.

In Svalbard, polar bears were observed to kill white-beaked dolphins during spring, when the dolphins were trapped in the sea ice. The bears then proceeded to cache the carcasses, which remained and were eaten during the ice-free summer and autumn.

Reproduction and Lifecycle

Cubs are born helpless and typically nurse for two and a half years.

Courtship and mating take place on the sea ice in April and May, when polar bears congregate in the best seal hunting areas. A male may follow the tracks of a breeding female for 100 km (60 mi) or more, and after finding her engage in intense fighting with other males over mating rights, fights that often result in scars and broken teeth. Polar bears have a generally polygynous mating system; recent genetic testing of mothers and cubs, however, has uncovered cases of litters in which cubs have different fathers. Partners stay together and mate repeatedly for an entire week; the mating ritual induces ovulation in the female.

After mating, the fertilized egg remains in a suspended state until August or September. During these four months, the pregnant female eats prodigious amounts of food, gaining at least 200 kg (440 lb) and often more than doubling her body weight.

Maternity Denning and Early Life

Mother and cub on Svalbard.

When the ice floes are at their minimum in the fall, ending the possibility of hunting, each pregnant female digs a *maternity den* consisting of a narrow entrance tunnel leading to one to three chambers. Most maternity dens are in snowdrifts, but may also be made underground in permafrost if it is not sufficiently cold yet for snow. In most subpopulations, maternity dens are situated on land a few kilometers from the coast, and the individuals in a subpopulation tend to reuse the same denning areas each year. The polar bears that do not den on land make their dens on the sea ice. In the den, she enters a dormant state similar to hibernation. This hibernation-like state does not consist of continuous sleeping; however, the bear's heart rate slows from 46 to 27 beats per minute. Her body temperature does not decrease during this period as it would for a typical mammal in hibernation.

Between November and February, cubs are born blind, covered with a light down fur, and weighing less than 0.9 kg (2.0 lb), but in captivity they might be delivered in the earlier months. The earliest recorded birth of polar bears in captivity was on 11 October 2011 in the Toronto Zoo. On average, each litter has two cubs. The family remains in the den until mid-February to mid-April, with the mother maintaining her fast while nursing her cubs on a fat-rich milk. By the time the mother breaks open the entrance to the den, her cubs weigh about 10 to 15 kilograms (22 to 33 lb). For about 12 to 15 days, the family spends time outside the den while remaining in its vicinity, the mother grazing on vegetation while the cubs become used to walking and playing. Then they begin the long walk from the denning area to the sea ice, where the mother can once again catch seals. Depending on the timing of ice-floe breakup in the fall, she may have fasted for up to eight months. During this time, cubs playfully imitate the mother's hunting methods in preparation for later life.

Female polar bears are noted for both their affection towards their offspring, and their valor in protecting them. Multiple cases of adoption of wild cubs have been confirmed by genetic testing. Adult bears of either gender occasionally kill and eat polar bear cubs. As of 2006, in Alaska, 42% of cubs were reaching 12 months of age, down from 65% in 1991. In most areas, cubs are weaned at two and a half years of age, when the mother chases them away or abandons them. The Western Hudson Bay subpopulation is unusual in that its female polar bears sometimes wean their cubs at only one and a half years. This was the case for 40% of cubs there in the early 1980s; however by the 1990s, fewer than 20% of cubs were weaned this young. After the mother leaves, sibling cubs sometimes travel and share food together for weeks or months.

Later Life

Females begin to breed at the age of four years in most areas, and five years in the Beaufort Sea area. Males usually reach sexual maturity at six years; however, as competition for females is fierce, many do not breed until the age of eight or ten. A study in Hudson Bay indicated that both the reproductive success and the maternal weight of females peaked in their mid-teens.

Polar bears appear to be less affected by infectious diseases and parasites than most terrestrial mammals. Polar bears are especially susceptible to *Trichinella*, a parasitic roundworm they contract through cannibalism, although infections are usually not fatal. Only one case of a polar bear with rabies has been documented, even though polar bears frequently interact with Arctic foxes, which often carry rabies. Bacterial leptospirosis and *Morbillivirus* have been recorded. Polar bears sometimes have problems with various skin diseases that may be caused by mites or other parasites.

Life Expectancy

Polar bears rarely live beyond 25 years. The oldest wild bears on record died at age 32, whereas the oldest captive was a female who died in 1991, age 43. The causes of death in wild adult polar bears are poorly understood, as carcasses are rarely found in the species's frigid habitat. In the wild, old polar bears eventually become too weak to catch food, and gradually starve to death. Polar bears injured in fights or accidents may either die from their injuries or become unable to hunt effective-ly, leading to starvation.

Ecological Role

The polar bear is the apex predator within its range, and is a keystone species for the Arctic. Sev-eral animal species, particularly Arctic foxes (*Vulpes lagopus*) and glaucous gulls (*Larus hyperbo-reus*), routinely scavenge polar bear kills.

The relationship between ringed seals and polar bears is so close that the abundance of ringed seals in some areas appears to regulate the density of polar bears, while polar bear predation in turn regulates density and reproductive success of ringed seals. The evolutionary pressure of polar bear predation on seals probably accounts for some significant differences between Arctic and Antarctic seals. Compared to the Antarctic, where there is no major surface pred-ator, Arctic seals use more breathing holes per individual, appear more restless when hauled out on the ice, and rarely defecate on the ice. The baby fur of most Arctic seal species is white, presumably to provide camouflage from predators, whereas Antarctic seals all have dark fur at birth.

Brown bears tend to dominate polar bears in disputes over carcasses, and dead polar bear cubs have been found in brown bear dens. Wolves are rarely encountered by polar bears, though there are two records of Arctic wolf (*Canis lupus arctos*) packs killing polar bear cubs. Adult polar bears are occasionally vulnerable to predation by orcas (*Orcinus orca*) while swimming, but they are rarely reported as taken and bears are likely to avoid entering the water if possible if they detect an orca pod in the area. The melting sea ice in the Arctic may be causing an increase of orcas in the Arctic sea, which may increase the risk of predation on polar bears but also may benefit the bears by providing more whale carcasses that they can scavenge. The remains of polar bears have found in the stomachs of large Greenland sharks (*Somniosus microcephalus*), although it certainly cannot be ruled out that the bears were merely scavenged by this slow-moving, unusual shark. A rather unlikely killer of a grown polar bear has reportedly included a wolverine (*Gulo gulo*), anec-dotely reported to have suffocated a bear in a zoo with a bite to the throat during a conflict. This report may well be dubious, however. Polar bears are sometimes the host of arctic mites such as *Alaskozetes antarcticus*.

Long-distance Swimming and Diving

Researchers tracked 52 sows in the southern Beaufort Sea off Alaska with GPS system collars; no boars were involved in the study due to males' necks being too thick for the GPS-equipped collars. Fifty long-distance swims were recorded; the longest at 354 kilometres (220 mi), with an average of 155 kilometres (96 mi). The length of these swims ranged from most of a day to ten days. Ten of the sows had a cub swim with them and after a year, six cubs survived. The study did not determine if the others lost their cubs before, during, or some time after their long swims. Researchers do not know whether or not this is a new behaviour; before polar ice shrinkage, they opined that there was probably neither the need nor opportunity to swim such long distances.

The polar bear may swim underwater for up to three minutes to approach seals on shore or on ice floes.

Hunting

Indigenous People

Skins of hunted polar bears.

Polar bears have long provided important raw materials for Arctic peoples, including the Inuit, Yupik, Chukchi, Nenets, Russian Pomors and others. Hunters commonly used teams of dogs to distract the bear, allowing the hunter to spear the bear or shoot it with arrows at closer range. Almost all parts of captured animals had a use. The fur was used in particular to make trousers and, by the Nenets, to make galoshes-like outer footwear called *tobok*; the meat is edible, despite some risk of trichinosis; the fat was used in food and as a fuel for lighting homes, alongside seal and whale blubber; sinews were used as thread for sewing clothes; the gallbladder and sometimes heart were dried and powdered for medicinal purposes; the large canine teeth were highly valued as talismans. Only the liver was not used, as its high concentration of vitamin A is poisonous. Hunters make sure to either toss the liver into the sea or bury it in order to spare their dogs from potential poisoning. Traditional subsistence hunting was on a small enough scale to not significantly affect polar bear populations, mostly because of the sparseness of the human population in polar bear habitat.

Commercial Harvest

In Russia, polar bear furs were already being commercially traded in the 14th century, though it was of low value compared to Arctic fox or even reindeer fur. The growth of the human population in the Eurasian Arctic in the 16th and 17th century, together with the advent of firearms and increasing trade, dramatically increased the harvest of polar bears. However, since polar bear fur has always played a marginal commercial role, data on the historical harvest is fragmentary. It is known, for example, that already in the winter of 1784/1785 Russian Pomors on Spitsbergen harvested 150 polar bears in Magdalenefjorden. In the early 20th century, Norwegian hunters were harvesting 300 bears per year at the same location. Estimates of total historical harvest suggest that from the beginning of the 18th century, roughly 400 to 500 animals were being harvested annually in northern Eurasia, reaching a peak of 1,300 to 1,500 animals in the early 20th century, and falling off as the numbers began dwindling.

In the first half of the 20th century, mechanized and overpoweringly efficient methods of hunting and trapping came into use in North America as well. Polar bears were chased from snowmobiles, icebreakers, and airplanes, the latter practice described in a 1965 *New York Times* editorial as being "about as sporting as machine gunning a cow." Norwegians used "self-killing guns", comprising a loaded rifle in a baited box that was placed at the level of a bear's head, and which fired when the string attached to the bait was pulled. The numbers taken grew rapidly in the 1960s, peaking around 1968 with a global total of 1,250 bears that year.

Contemporary Regulations

Road sign warning about the presence of polar bears.

Concerns over the future survival of the species led to the development of national regulations on polar bear hunting, beginning in the mid-1950s. The Soviet Union banned all hunting in 1956. Canada began imposing hunting quotas in 1968. Norway passed a series of increasingly strict regulations from 1965 to 1973, and has completely banned hunting since then. The United States began regulating hunting in 1971 and adopted the Marine Mammal Protection Act in 1972. In 1973, the International Agreement on the Conservation of Polar Bears was signed by all five nations whose territory is inhabited by polar bears: Canada, Denmark, Norway, the Soviet Union, and the United States. Member countries agreed to place restrictions on recreational and commercial hunting, ban hunting from aircraft and icebreakers, and conduct further research. The treaty allows hunting "by local people using traditional methods". Norway is the only country of the five in which all harvest of polar bears is banned. The agreement was a rare case of international cooperation during the Cold War. Biologist Ian Stirling commented, "For many years, the conservation of polar bears was the only subject in the entire Arctic that nations from both sides of the Iron Curtain could agree upon sufficiently to sign an agreement. Such was the intensity of human fascination with this magnificent predator, the only marine bear."

Agreements have been made between countries to co-manage their shared polar bear subpopulations. After several years of negotiations, Russia and the United States signed an agreement in October 2000 to jointly set quotas for indigenous subsistence hunting in Alaska and Chukotka. The treaty was ratified in October 2007. In September 2015, the polar bear range states agreed upon a "circumpolar action plan" describing their conservation strategy for polar bears.

Although the United States government has proposed that polar bears be transferred to Appendix I of CITES, which would ban all international trade in polar bear parts, polar bears currently remain listed under Appendix II. This decision was approved of by members of the IUCN and TRAFFIC, who determined that such an uplisting was unlikely to confer a conservation benefit.

Polar bears were designated "Not at Risk" in April 1986 and uplisted to "Special Concern" in April 1991. This status was re-evaluated and confirmed in April 1999, November 2002, and April 2008. Polar bears continue to be listed as a species of special concern in Canada because of their sensitivity to overharvest and because of an expected range contraction caused by loss of Arctic sea ice.

Dogsleds are used for recreational hunting of polar bears in Canada.

More than 600 bears are killed per year by humans across Canada, a rate calculated by scientists to be unsustainable for some areas, notably Baffin Bay. Canada has allowed sport hunters accompanied by local guides and dog-sled teams since 1970, but the practice was not common until the 1980s. The guiding of sport hunters provides meaningful employment and an important source of income for northern communities in which economic opportunities are few. Sport hunting can bring CDN$20,000 to $35,000 per bear into northern communities, which until recently has been mostly from American hunters.

The territory of Nunavut accounts for the location 80% of annual kills in Canada. In 2005, the government of Nunavut increased the quota from 400 to 518 bears, despite protests from the IUCN Polar Bear Specialist Group. In two areas where harvest levels have been increased based on increased sightings, science-based studies have indicated declining populations, and a third area is considered data-deficient. While most of that quota is hunted by the indigenous Inuit people, a growing share is sold to recreational hunters. (0.8% in the 1970s, 7.1% in the 1980s, and 14.6% in the 1990s) Nunavut polar bear biologist, Mitchell Taylor, who was formerly responsible for polar bear conservation in the territory, has insisted that bear numbers are being sustained under current hunting limits. In 2010, the 2005 increase was partially reversed. Government of Nunavut officials announced that the polar bear quota for the Baffin Bay region would be gradually reduced from 105 per year to 65 by the year 2013. The Government of the Northwest Territories maintain their own quota of 72 to 103 bears within the Inuvialuit communities of which some are set aside for sports hunters. Environment Canada also banned the export from Canada of fur, claws, skulls and other products from polar bears harvested in Baffin Bay as of 1 January 2010.

Because of the way polar bear hunting quotas are managed in Canada, attempts to discourage sport hunting would actually increase the number of bears killed in the short term. Canada allocates a certain number of permits each year to sport and subsistence hunting, and those that are not used for sport hunting are re-allocated to indigenous subsistence hunting. Whereas northern communities kill all the polar bears they are permitted to take each year, only half of sport hunters with permits actually manage to kill a polar bear. If a sport hunter does not kill a polar bear before his or her permit expires, the permit cannot be transferred to another hunter.

In August 2011, Environment Canada published a national polar bear conservation strategy.

In Greenland, hunting restrictions were first introduced in 1994 and expanded by executive order in 2005. Until 2005 Greenland placed no limit on hunting by indigenous people. However, in 2006 it imposed a limit of 150, while also allowed recreational hunting for the first time. Other provisions included year-round protection of cubs and mothers, restrictions on weapons used and various administrative requirements to catalogue kills.

Polar bear were hunted heavily in Svalbard, Norway throughout the 19th century and to as recently as 1973, when the conservation treaty was signed. 900 bears a year were harvested in the 1920s and after World War II, there were as many as 400–500 harvested annually. Some regulations of hunting did exist. In 1927, poisoning was outlawed while in 1939, certain denning sights were declared off limits. The killing of females and cubs was made illegal in 1965. Killing of polar bears decreased somewhat 25–30 years before the treaty. Despite this, the polar bear population continued to decline and by 1973, only around 1000 bears were left in Svalbard. Only with the passage of the treaty did they begin to recover.

The Soviet Union banned the harvest of polar bears in 1956; however, poaching continued and is estimated to pose a serious threat to the polar bear population. In recent years, polar bears have approached coastal villages in Chukotka more frequently due to the shrinking of the sea ice, endangering humans and raising concerns that illegal hunting would become even more prevalent. In 2007, the Russian government made subsistence hunting legal for indigenous Chukotkan peoples only, a move supported by Russia's most prominent bear researchers and the World Wide Fund for Nature as a means to curb poaching.

Polar bears are currently listed as "Rare", of "Uncertain Status", or "Rehabilitated and rehabilitating" in the Red Data Book of Russia, depending on population. In 2010, the Ministry of Natural Resources and Environment published a strategy for polar bear conservation in Russia.

The Marine Mammal Protection Act of 1972 afforded polar bears some protection in the United States. It banned hunting (except by indigenous subsistence hunters), banned importing of polar bear parts (except polar bear pelts taken legally in Canada), and banned the harassment of polar bears. On 15 May 2008, the United States Department of the Interior listed the polar bear as a threatened species under the Endangered Species Act, citing the melting of Arctic sea ice as the primary threat to the polar bear. It banned all importing of polar bear trophies. Importing products made from polar bears had been prohibited from 1972 to 1994 under the Marine Mammal Protection Act, and restricted between 1994 and 2008. Under those restrictions, permits from the United States Fish and Wildlife Service were required to import sport-hunted polar bear trophies taken in hunting expeditions in Canada. The permit process required that the bear be taken from an area with quotas based on sound management principles. Since 1994, hundreds of sport-hunted polar bear trophies have been imported into the U.S. In 2015, the U.S. Fish and Wildlife Service published a draft conservation management plan for polar bears to improve their status under the Endangered Species Act and the Marine Mammal Protection Act.

Conservation Status, Threats and Controversies

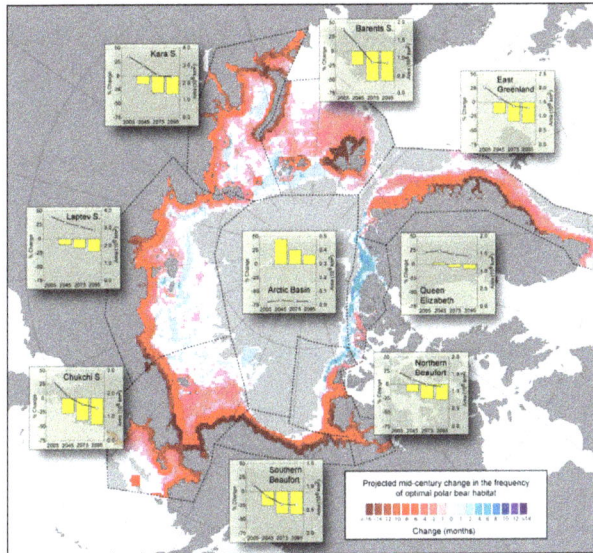

Map from the U.S. Geological Survey shows projected changes in polar
bear habitat from 2001 to 2010 and 2041 to 2050. Red areas indicate loss of
optimal polar bear habitat; blue areas indicate gain.

Polar bear population sizes and trends are difficult to estimate accurately because they occupy remote home ranges and exist at low population densities. Polar bear fieldwork can also be hazardous to researchers. As of 2015, the International Union for Conservation of Nature (IUCN) reports that the global population of polar bears is 22,000 to 31,000, and the current population trend is unknown. Nevertheless, polar bears are listed as "Vulnerable" under criterion A3c, which indicates an expected population decrease of ≥30% over the next three generations (~34.5 years) due to "decline in area of occupancy, extent of occurrence and/or quality of habitat". Risks to the polar bear include climate change, pollution in the form of toxic contaminants, conflicts with shipping, oil and gas exploration and development, and human-bear interactions including harvesting and possible stresses from recreational polar-bear watching.

According to the World Wildlife Fund, the polar bear is important as an indicator of Arctic ecosystem health. Polar bears are studied to gain understanding of what is happening throughout the Arctic, because at-risk polar bears are often a sign of something wrong with the Arctic marine ecosystem.

Climate Change

The International Union for Conservation of Nature, Arctic Climate Impact Assessment, United States Geological Survey and many leading polar bear biologists have expressed grave concerns about the impact of climate change, including the belief that the current warming trend imperils the survival of the polar bear.

The key danger posed by climate change is malnutrition or starvation due to habitat loss. Polar bears hunt seals from a platform of sea ice. Rising temperatures cause the sea ice to melt earlier in the year, driving the bears to shore before they have built sufficient fat reserves to survive the period of scarce food in the late summer and early fall. Reduction in sea-ice cover also forces bears to swim longer distances, which further depletes their energy stores and occasionally leads to drowning. Thinner sea ice tends to deform more easily, which appears to make it more difficult for polar bears to access seals. Insufficient nourishment leads to lower reproductive rates in adult females and lower survival rates in cubs and juvenile bears, in addition to poorer body condition in bears of all ages.

Mothers and cubs have high nutritional requirements,
which are not met if the seal-hunting season is too short.

In addition to creating nutritional stress, a warming climate is expected to affect various other aspects of polar bear life: Changes in sea ice affect the ability of pregnant females to build suitable maternity dens. As the distance increases between the pack ice and the coast, females must swim longer distances to reach favoured denning areas on land. Thawing of permafrost would affect the bears who traditionally den underground, and warm winters could result in den roofs collapsing or having reduced insulative value. For the polar bears that currently den on multi-year ice, increased ice mobility may result in longer distances for mothers and young cubs to walk when they return to seal-hunting areas in the spring. Disease-causing bacteria and parasites would flourish more readily in a warmer climate.

Problematic interactions between polar bears and humans, such as foraging by bears in garbage dumps, have historically been more prevalent in years when ice-floe breakup occurred early and local polar bears were relatively thin. Increased human-bear interactions, including fatal attacks on humans, are likely to increase as the sea ice shrinks and hungry bears try to find food on land.

Starving polar bear near Svalbard.

The effects of climate change are most profound in the southern part of the polar bear's range, and this is indeed where significant degradation of local populations has been observed. The Western Hudson Bay subpopulation, in a southern part of the range, also happens to be one of the best-studied polar bear subpopulations. This subpopulation feeds heavily on ringed seals in late spring, when newly weaned and easily hunted seal pups are abundant. The late spring hunting season ends for polar bears when the ice begins to melt and break up, and they fast or eat little during the summer until the sea freezes again.

Due to warming air temperatures, ice-floe breakup in western Hudson Bay is currently occurring three weeks earlier than it did 30 years ago, reducing the duration of the polar bear feeding season. The body condition of polar bears has declined during this period; the average weight of lone (and likely pregnant) female polar bears was approximately 290 kg (640 lb) in 1980 and 230 kg (510 lb) in 2004. Between 1987 and 2004, the Western Hudson Bay population declined by 22%, although the population was listed as "stable" as of 2017. As the climate change melts sea ice, the U.S. Geological Survey projects that two-thirds of polar bears will disappear by 2050.

In Alaska, the effects of sea ice shrinkage have contributed to higher mortality rates in polar bear cubs, and have led to changes in the denning locations of pregnant females. In recent years, polar bears in the Arctic have undertaken longer than usual swims to find prey, possibly resulting in four recorded drownings in the unusually large ice pack regression of 2005.

A new development is that polar bears have begun ranging to new territory. While not unheard of but still uncommon, polar bears have been sighted increasingly in larger numbers ashore, staying on the mainland for longer periods of time during the summer months, particularly in North Canada, traveling farther inland. This may cause an increased reliance on terrestrial diets, such as goose eggs, waterfowl and caribou, as well as increased human–bear conflict.

Pollution

Polar bears accumulate high levels of persistent organic pollutants such as polychlorinated biphenyl (PCBs) and chlorinated pesticides. Due to their position at the top of the ecological pyramid, with a diet heavy in blubber in which halocarbons concentrate, their bodies are among the most contaminated of Arctic mammals. Halocarbons are known to be toxic to other animals, because they mimic hormone chemistry, and biomarkers such as immunoglobulin G and retinol suggest similar effects on polar bears. PCBs have received the most study, and they have been associated with birth defects and immune system deficiency.

Many chemicals, such as PCBs and DDT, have been internationally banned due to the recognition of their harm on the environment. Their concentrations in polar bear tissues continued to rise for decades after being banned as these chemicals spread through the food chain. Since then, the trend seems to have discontinued, with tissue concentrations of PCBs declining between studies performed from 1989 to 1993 and studies performed from 1996 to 2002. During the same time periods, DDT was notably lower in the Western Hudson Bay population only.

Oil and Gas Development

Oil and gas development in polar bear habitat can affect the bears in a variety of ways. An oil spill in the Arctic would most likely concentrate in the areas where polar bears and their prey are also concentrated, such as sea ice leads. Because polar bears rely partly on their fur for insulation and soiling of the fur by oil reduces its insulative value, oil spills put bears at risk of dying from hypothermia. Polar bears exposed to oil spill conditions have been observed to lick the oil from their fur, leading to fatal kidney failure. Maternity dens, used by pregnant females and by females with infants, can also be disturbed by nearby oil exploration and development. Disturbance of these sensitive sites may trigger the mother to abandon her den prematurely, or abandon her litter altogether.

Predictions

Steven Amstrup and other U.S. Geological Survey scientists have predicted two-thirds of the world's polar bears may disappear by 2050, based on moderate projections for the shrinking of summer sea ice caused by climate change, though the validity of this study has been debated. The bears could disappear from Europe, Asia, and Alaska, and be depleted from the Canadian Arctic Archipelago and areas off the northern Greenland coast. By 2080, they could disappear from Greenland entirely and from the northern Canadian coast, leaving only dwindling numbers in the interior Arctic Archipelago. However, in the short term, some polar bear populations in historically colder regions of the Arctic may temporarily benefit from a milder climate, as multiyear ice that is too thick for seals to create breathing holes is replaced by thinner annual ice.

Polar bears diverged from brown bears 400,000–600,000 years ago and have survived past periods of climate fluctuation. It has been claimed that polar bears will be able to adapt to terrestrial food sources as the sea ice they use to hunt seals disappears. However, most polar bear biologists think that polar bears will be unable to completely offset the loss of calorie-rich seal blubber with terrestrial foods, and that they will be outcompeted by brown bears in this terrestrial niche, ultimately leading to a population decline.

7

Threats to Marine Mammals and their Protection

There are numerous human activities which threaten marine mammals. A few of these are ship strikes, acoustic pollution, open net fishing, commercial hunting, oil spills, etc. Some of the steps taken to protect them are the Marine Mammal Protection Act and the establishment of marine protected areas. All these diverse threats to marine mammals and the measures to protect them have been carefully analyzed in this chapter.

MODERN THREATS TO MARINE ANIMALS

Ship Strikes

Ship strikes are the number one culprit in the disappearance of the North Atlantic Right Whale. The Right Whale is only one of many marine mammal species that is at risk of being hit by high-speed ships. It is estimated that over 90 manatees die prematurely due to ship strikes every year in Florida.

Acoustic Pollution

Whale and dolphins rely on acoustics to navigate their way through the ocean. These mammals rely on sound for communication, mating, foraging, and migration. The addition of loud noises from ships, sonar, drilling rigs, and other human sources can distort messages sent by marine mammals. Scientists believe that acoustic noise pollution prevents these mammals from being able to detect approaching ships or fishing nets, adding to the risk of being killed.

Open Net Fishing

Trawling and gillnetting are two types of fishing methods that involve running extremely long lengths of fishing net through open water. These nets are usually left unattended to and have been devastating to sharks, dolphins, sea turtles, and whales that accidentally get caught in them. It is estimated that these terrible fishing methods are responsible for the overfishing of 90 percent of the world's large predatory fish. These methods are also the number one cause of death for the Maui's dolphin.

Oil Spills

Oil spills have both short term and long term effects on marine mammals. Animals like seals and otters that are exposed to oil will automatically try and lick oil off their fur. Toxic chemicals in oil causes serious damage to the digestive system and internal organs. Thick, sticky oil has also been shown to clog the blowholes of dolphins and whales, causing them to suffocate. In the long term, exposure to petroleum can cause reproductive damage, making it difficult for populations to re-populate in the wake of oil spills. Otter populations are only just starting to come back after the Exxon-Valdez spill.

Agricultural Runoff

Fertilizer and pesticide runoff from farms across the world pose a serious threat to marine mammal's ecosystems. Runoff is especially problematic in costal regions where the excess nitrogen from fertilizer can spawn massive algae blooms that deplete water of oxygen creating a "dead zone" for fish. While small fish may be most effected, the chemicals from agricultural runoff bioaccumulate up the food chain. In some places, researchers have recovered beached Beluga whales that are so saturated with toxic chemicals from agricultural chemicals they must be handled like toxic waste.

Commercial Hunting

Whaling, shark, and dolphin hunting are among some of the most imminent threats to marine mammals. Slaughtering animals for sport or to sell as specialty foods is a practice that has greatly reduced the number of large marine mammals across the world's oceans. During the 2013 to 2014 season, Japan's Taiji dolphin hunt was responsible for the deaths of more than 800 dolphins. Shark finning is another inhumane, senseless practice that slaughters on average 100 million sharks a year.

Climate Change

Climate change is already making a huge impact on the lives of marine mammals, especially in the Arctic. Because their habitat is literally melting away, mammals like the ringed seal must find new places to give birth and raise their young. As the ocean absorbs more carbon dioxide from the atmosphere, global waters are becoming more acidic. High pH levels hinder crustacean's ability to form shells and can harm fish larvae. This effect ripples up the food chain, leaving larger marine mammals without food sources, causing major damage to marine ecosystems across the entire world.

Entertainment and Captivity

Dolphins and whales are amazing creatures to watch, but capturing them from the wild and forc-ing them to live in captivity for entertainment purposes has a serious impact on natural marine ecosystems. Taking a single dolphin or whale from a pod can completely disrupt the pod's function. Over 145 orca whales have been forced into captivity over the lifespan of the whale captivity business, while nearly 90 percent of these whales died. In the wild, orcas can live to be 90 years old. However, the average orca in captivity only lives four or five years.

Tourism

The sale of wild animal parts is illegal! However, novelty souvenirs like shark teeth, real tortoise shell jewelry, coral necklaces, walrus tusks, and fur from seals are a HUGE problem in global tourism. Although these products might be illegal to bring in and out of the United States, that does not mean that helpless marine animals are not slaughtered in the name of novelty trinkets.

Habitat Loss

By and large, loss of habitat is the BIGGEST threat to the livelihood of marine mammals. Habitat loss can occur as the result of pollution, changes in ecosystems, ship traffic, and a number of other human-related problems. Only 0.6 percent of the ocean is protected under conservation laws, meaning there is a whole lot of open water that is fair game.

SEAL HUNTING

Seal hunting, or sealing, is the personal or commercial hunting of seals. Seal hunting is currently practiced in nine countries and one region of Denmark: United States, Canada, Namibia, Iceland, Norway, Russia, Finland, Sweden, and Greenland. Most of the world's seal hunting takes place in Canada and Greenland.

The Canadian Department of Fisheries and Oceans (DFO) regulates the seal hunt in Canada. It sets quotas (total allowable catch – TAC), monitors the hunt, studies the seal population, works with the Canadian Sealers' Association to train sealers on new regulations, and promotes sealing through its website and spokespeople. The DFO set harvest quotas of over 90,000 seals in 2007; 275,000 in 2008; 280,000 in 2009; and 330,000 in 2010. The actual kills in recent years have been less than the quotas: 82,800 in 2007; 217,800 in 2008; 72,400 in 2009; and 67,000 in 2010.

In 2007, Norway claimed that 29,000 harp seals were killed, Russia claimed that 5,479 seals were killed, and Greenland claimed that 90,000 seals were killed in their respective seal hunts.

Harp seal populations in the northwest Atlantic declined to approximately 2 million in the late 1960s as a result of Canada's annual kill rates, which averaged to over 291,000 from 1952 to 1970. Conservationists demanded reduced rates of killing and stronger regulations to avert the extinction of the harp seal. In 1971, the Canadian government responded by instituting a quota system. The system was competitive, with each boat catching as many seals as it could before the hunt closed, which the Department of Fisheries and Oceans did when they knew that year's quota had been reached. Because it was thought that the competitive element might cause sealers to cut corners, new regulations were introduced that limited the catch to 400 seals per day, and 2000 per boat total. A 2007 population survey conducted by the DFO estimated the population at 5.5 million.

It is illegal in Canada to hunt newborn harp seals (whitecoats) and young hooded seals (bluebacks). When the seal pups begin to molt their downy white fur at the age of 12–14 days, they are called "ragged-jacket" and can be commercially hunted. After molting, the seals are called "beaters", named for the way they beat the water with their flippers. The hunt remains highly controversial, attracting significant media coverage and protests each year. Images from past hunts have become iconic symbols for conservation, animal welfare, and animal rights advocates. In 2009, Russia banned the hunting of harp seals less than one year old.

The term seal is used to refer to a diverse group of animals. In science, they are grouped together in the Pinnipeds, which also includes the walrus, not popularly thought of as a seal, and not considered here. The two main families of seals are the Otariidae (the eared seals; includes sea lions, and fur seals), and Phocidae (the earless seals); animals in the family Phocidae are sometimes referred to as hair seals, and are much more adept in the water than the eared seals, though they have a more difficult time getting around on land. The fur seal yields a valuable fur; the hair seal has no fur, but oil can be obtained from its fat and leather from its hide. Seals have been used for their pelts, their flesh, and their fat, which was often used as lamp fuel, lubricants, cooking oil, a constituent of soap, the liquid base for red ochre paint, and for processing materials such as leather and jute.

Traditional Seal Hunting

Archeological evidence indicates the Native Americans and First Nations People in Canada have been hunting seals for at least 4,000 years. Traditionally, when an Inuit boy killed his first seal or caribou, a feast was held. The meat was an important source of fat, protein, vitamin A, vitamin B12 and iron, and the pelts were prized for their warmth. The Inuit diet is rich in fish, whale, and seal.

Inuit seal hunting.

There were approximately 150,000 circumpolar Inuit in 2005 in Greenland, Alaska, Russia, and Canada. According to Kirt Ejesiak, former secretary and chief of staff to then-Premier of Nunavut, Paul Okalik and the first Inuk from Nunavut to attend Harvard, for the c. 46,000 Canadian Inuit, the seal was not "just a source of cash through fur sales, but the keystone of their culture. Although Inuit harvest and hunt many species that inhabit the desert tundra and ice platforms, the seal is their mainstay. The Inuktitut vocabulary designates specific objects made from seal bone, sinew, fat and fur used as tools, games, thread, cords, fuel, clothing, boats, and tents. There are also words referring to seasons, topography, place names, legends, and kinship relationships based on the seal. One region of Canada's north is inhabited by the Netsilingmiut, or "people of the seal." Wenzel's "scholarly examination" of "the impact of the animal rights movement upon the culture and economy of the Canadian Inuit" was among the first to reveal how animal rights groups, "well-meaning people in the dominant society through misunderstanding and ignorance can inflict destruction" on a vulnerable minority.

Inuit seal hunting accounts for the majority of the seal hunt, but just three percent of the hunt in southern Canada; it is excluded from the European Commission's call in 2006 for a ban on the import, export and sale of all harp and hooded seal products. Ringed seals were once the main staple for food, and have been used for clothing, boots, fuel for lamps, as delicacy, containers, igloo windows, and in harnesses for huskies. Though no longer used to this extent, ringed seals are still an important food and clothing source for the people of Nunavut. Called nayiq by the Central Alaskan Yup'ik people, the ringed seal is also hunted and eaten in Alaska. Seals were also hunted in northwest Europe and the Baltic Sea more than 10,000 years ago.

Early Modern Era

Seal hunting in Reimerswaal.

The first commercial hunting of seals by Europeans, is said to have occurred in 1515, when a cargo of fur seal skins from Uruguay was sent to Spain for sale in the markets of Seville.

Newfoundland

Newfoundland and Labrador and the Gulf of St. Lawrence were the first regions to experience large scale sealing. Migratory fishermen began the hunting from as early as the 1500s. Large-scale commercial seal hunting became an annual event starting in 1723 and expanded rapidly near the turn of the 18th century. Initially, the method used was to ensnare the migrating seals in nets anchored to shore installations, known as the 'landsman seal fishery'. The hunt was mainly for the procurement of seal meat as a form of sustenance for the settlements in the area, rather than for commercial gain.

From the early 18th century English hunters began to range further afield - 1723 marked the first time that hunters armed with firearms ventured forth in boats to increase their haul. This soon became a sophisticated commercial operation; the seals were transported back to England, where the seal's meat, fur, and oil were sold separately. From 1749, the import of seal oil to England was being recorded annually, and was used as lighting oil, for cooking, in the manufacture of soap and for the treating of leather.

South Atlantic

It was in the South Seas that sealing became a major enterprise from the late 18th century. Samuel Enderby, along with Alexander Champion and John St Barbe organized the first commercial expedition to the South Atlantic Ocean in 1776, initially with the primary aim of whaling, although sealing began to play a prominent part in the operation as well. More expeditions were sent in 1777 and 1778 before political and economic troubles hampered the trade for some time.

On 1 September 1788, the 270 ton ship Emilia, owned by Samuel Enderby & Sons and commanded by Captain James Shields, departed London. The ship went west around Cape Horn into the Pacific Ocean to become the first ship of any nation to conduct operations in the Southern Ocean. Emilia returned to London on 12 March 1790 with a cargo of 139 tons of sperm oil.

By 1784, the British had fifteen ships in the southern fishery, all from London. By 1790 this port alone had sixty vessels employed in the trade. Between 1793 and 1799 there was an average of sixty vessels in the trade. The average increased to seventy-two in the years between 1800 and 1809.

The sealing industry extended further south to the South Georgia island, first mapped by Captain James Cook in HMS Resolution on 17 January 1775. During the late 18th century and throughout the 19th century, South Georgia was inhabited by English and Yankee sealers, who used to live there for considerable periods of time and sometimes overwintered. In 1778, English sealers brought back from the Island of South Georgia and the Magellan Strait area as many as 40,000 seal skins and 2,800 tons of elephant seal oil. More fur seals from the island were taken in 1786 by the English sealing vessel Lord Hawkesbury, and by 1791, 102 vessels, manned by 3000 sealers, were hunting seals south of the equator. The first commercial visit to the South Sandwich Islands was made in 1816 by another English ship, the Ann.

The sealers pursued their trade in a most unsustainable manner, promptly reducing the fur seal population to near extermination. As a result, sealing activities on South Georgia had three marked peaks in 1786–1802, 1814–23, and 1869–1913 respectively, decreasing in between and gradually shifting to elephant seals taken for oil.

Pacific

Commercial sealing in Australasia appears to have started with the London-based Massachusetts-born Eber Bunker, master of the William and Ann, who announced his intention in November 1791 to visit and hunt in New Zealand's Dusky Sound.Captain William Raven of the Britannia stationed a party at Dusky from 1792 to 1793, but the discovery in 1798-1799 of Bass Strait, between mainland Australia and Van Diemen's Land (later known as Tasmania) saw the sealers' focus shift there in 1798, when a gang including Daniel Cooper landed from the Nautilus on Cape Barren Island.

With Bass Strait over-exploited by 1802, commercial attention returned to southern New Zealand waters, where Stewart Island/Rakiura and Foveaux Strait were explored, exploited and charted from 1803 to 1804. Thereafter, the sealing-industry focus shifted to the sub-Antarctic Antipodes Islands, 1805–1807, the Auckland Islands from 1806, the southeast coast of New Zealand's South Island, Otago Harbour and Solander Island by 1809, before focusing further to the south at the newly discovered Campbell Island and Macquarie Island from 1810. During this period sealers were active on the southern coast of mainland Australia, for example at Kangaroo Island. This whole development has been called the first sealing boom; it sparked the Sealers' War in southern New Zealand.

By about 1815 sealing in the Pacific had faded in importance. A brief revival occurred from 1823, but this proved very short-lived. Although highly profitable at times and affording New South Wales one of its earliest trade staples, sealing's unregulated character saw its self-destruction. Notable traders from Britain and based in Australia included Simeon Lord, Henry Kable, James Underwood and Robert Campbell. Plummers of London and the Whitneys of New York also became involved.

By 1830 most Pacific seal stocks had been seriously depleted, and Lloyd's Register of Shipping only showed one full-time sealing vessel on its books. In the North Pacific, the later 1800s saw large harvests of fur seals. These harvests decreased along with fur-seal populations.

Industrial Era

Growing from the international Grand Banks fishery, the Newfoundland hunt initially used small schooners. Kill rates averaged 451,000 in the 1830s, and rose to 546,000 annually during the first half of the next decade, which led to a marked decline in the harp seal population that in turn adversely impacted profits in the sealing industry.

Sealing reached its peak in Newfoundland in the 1860s, with the introduction of more powerful and reliable steamships that were capable of much larger range and storing capacity. Annual catches exceeded the 400,000 mark from the 1870s and smaller sealers were steadily pushed out of the market.

The first modern sealing ship was the SS Bear, built in Dundee, Scotland in 1874 as a steamer for sealing. The ship was custom-built for sealing out of St. John's, Newfoundland, and was the most outstanding sealing vessel of her day and the lead ship in a new generation of sealers. Heavy-built with six inch (15.2 cm) thick wooden planks, Bear was rigged as a sailing barquentine but her main power was a steam engine designed to smash deep into ice packs to reach seal herds.

At the time of her arrival in St. John's, there were 300 vessels outfitted each season to hunt seals, but most were small schooners or old sailing barques. The new sealing ships represented by Bear radically transformed the Eastern North Atlantic seal fishery as they replaced the hundreds of smaller sealing vessels owned by merchants in outports around Newfoundland with large and expensive steamships owned by large British and Newfoundland companies based in St. John's. Owned at first by the Scottish firm W. Grieve and Sons, she was acquired in 1880 by R. Steele Junior.

Another famous sealing ship of the era was the Terra Nova, originally built in 1884 for the Dundee whaling and sealing fleet. She was ideally suited to the polar regions and worked for 10 years in the annual seal fishery in the Labrador Sea. Large and expensive ships required major capital investments from British and Newfoundland firms, and shifted the industry from merchants in small outports to companies based in St. John's, Newfoundland. By the late 19th century, the sealing industry in Newfoundland was second in importance only to cod fishing.

The seal hunt provided critical winter wages for fishermen, but was dangerous work marked by sealing disasters that claimed hundreds of lives, such as the 1914 Newfoundland Sealing Disaster involving the SS Southern Cross, the SS Newfoundland, and SS Stephano. The rugged hulls and experienced crews of Newfoundland sealing vessels often led sealers such as Bear and Terra Nova to be hired for arctic exploration and one sealer Algerine was hired to recover Titanic bodies in 1912. After World War II, the Newfoundland hunt was dominated by large Norwegian sealing vessels until the late 20th century, when the much diminished hunt shifted to smaller motor fishing vessels, based from outports around Newfoundland and Labrador. In 2007, the commercial seal hunt dividend contributed about $6 million to the Newfoundland GDP, a fraction of the industry's former importance.

Environmental Protection

The Russian Rurik, near Saint Paul Island in the Bering sea. Sealing in the Bering sea led to a diplomatic dispute between the US and Britain in the 1880s, which brought about the first legislative attempts to limit the environmental damage done by the sealers. Drawing by Louis Choris in 1817.

The end of the 19th century was marked by the Bering Sea Controversy between the United States and Great Britain over the management of fur seal harvests. In 1867 the United States government purchased from Russia all her territorial rights in Alaska and the adjacent islands, including the Pribilof Islands, the principal breeding-grounds of the seals in those seas. By Acts of Congress, the killing of seals was strictly regulated on the Pribiloff islands and in "the waters adjacent thereto". Beginning in about 1886, it became the practice of certain British and Canadian vessels to intercept passing seals in the open ocean (over three miles from any shore) and shoot them in the water (pelagic sealing). In the summer of 1886, three British Columbian sealers, the Carolena, Onward, and Thornton, were captured by an American revenue cutter, the Corwin, The United States claimed exclusive jurisdiction over the sealing industry in the Bering Sea; it also contended that the protection of the fur seal was an international duty, and should be secured by international arrangement. The British imperial government repudiated the claim, but was willing to negotiate on the question of international regulation.

While people living in Estonian coastal areas used to hunt seals, this has
changed a lot within the last half century. Within the last decade ecotourism
has also emerged, and nowadays seal watching tours are being organized.

The issue was taken to arbitration, which concluded in favour of Britain on all points in 1893. Since the decision was in favor of Great Britain, in accordance with the arbitration treaty the tribunal prescribed a series of regulations for preserving the seal herds which were to be binding upon and enforced by both powers. They limited pelagic sealing as to time, place, and manner by fixing a zone of 60 miles around the Pribilof Islands within which the seals were not to be molested at any time, and from May 1 to July 31 each year they were not to be pursued anywhere in Bering Sea. Only licensed sailing vessels were permitted to engage in fur sealing, and the use of firearms or explosives was prohibited.

This marked the first attempt at establishing regulations on the sealing industry for environmental purposes. However, these regulations failed because the mother seals fed outside the protected area and remained vulnerable. A joint commission of scientists from Britain and the United States further considered the problem, and came to the conclusion that the pelagic sealing needed to be curtailed. However further joint tribunals did not enact new legal restrictions, and then Japan also embarked upon pelagic sealing.

Finally, the North Pacific Fur Seal Convention severely curtailed the sealing industry. Signed on July 7, 1911 by the United States, Great Britain, Japan, and Russia, the treaty was designed to manage the commercial harvest of fur bearing mammals. It outlawed open-water seal hunting and acknowledged the United States' jurisdiction in managing the on-shore hunting of seals for commercial purposes. It was the first international treaty to address wildlife preservation issues.

The treaty was dissolved with the onset of hostilities between the signatories in World War II. However, the treaty set precedent for future national and international laws and treaties, including the Fur Seal Act of 1966 and the Marine Mammal Protection Act of 1972.

Today, commercial sealing is conducted by only five nations: Canada, Greenland, Namibia, Norway, and Russia. The United States, which had been heavily involved in the sealing industry, now maintains a complete ban on the commercial hunting of marine mammals, with the exception of indigenous peoples who are allowed to hunt a small number of seals each year.

Equipment and Method

A hakapik.

In the Canadian commercial seal hunt, the majority of the hunters initiate the kill using a firearm. Ninety percent of sealers on the ice floes of the Front (east of Newfoundland), where the majority of non-native seal hunting occurs, use firearms.

An older and more traditional method of killing seals is with a hakapik: a heavy wooden club with a hammer head and metal hook on the end. The hakapik is used because of its efficiency; the animal can be killed quickly without damage to its pelt. The hammer head is used to crush the seals' thin skulls, while the hook is used to move the carcasses. Canadian sealing regulations describe the dimensions of the clubs and the hakapiks, and caliber of the rifles and minimum bullet velocity, that can be used. They state: "Every person who strikes a seal with a club or hakapik shall strike the seal on the forehead until its skull has been crushed," and that "No person shall commence to skin or bleed a seal until the seal is dead," which occurs when it "has a glassy-eyed, staring appearance and exhibits no blinking reflex when its eye is touched while it is in a relaxed condition." Reportedly, in one study, three out of eight times, the animal was not rendered either dead or unconscious by shooting, and the hunters would then kill the seal using a hakapik or other club of a type that is sanctioned by the governing authority.

Modern Sealing

Products Made from Seals

Seal skins have been used by aboriginal people for millennia to make waterproof jackets and boots, and seal fur to make fur coats. Pelts account for over half the processed value of a seal, selling at over C$100 each as of 2006. According to Paul Christian Rieber, of GC Rieber AS, the difficult ice conditions and low quotas in 2006 resulted in less access to seal pelts, which caused the commodity price to be pushed up. One high-end fashion designer, Donatella Versace, has begun to use seal pelts, while others, such as Calvin Klein, Stella McCartney, Tommy Hilfiger, and Ralph Lauren, refrain from using any kind of fur.

A waistcoat made of seal fur.

Meat from a young harp seal.

Seal meat is an important source of food for residents of small coastal communities. Meat is sold to the Asian pet food market; in 2004, only Taiwan and South Korea purchased seal meat from Canada. The seal blubber is used to make seal oil, which is marketed as a fish oil supplement. In 2001, two percent of Canada's raw seal oil was processed and sold in Canadian health stores. There has been virtually no market for seal organs since 1998.

Sealing States

In 2005, three companies exported seal skin: Rieber in Norway, Atlantic Marine in Canada and Great Greenland in Greenland. Their clients were earlier French fashion houses and fur makers in Europe, but today the fur is mainly exported to Russia and China.

Canada

In Canada, the season for the commercial hunt of harp seal is from November 15 to May 15. While Inuit hunt seals commercially year-round, most sealing in southern Canada occurs in late March in the Gulf of St. Lawrence, and during the first or second week of April off Newfoundland, in an area known as the Front. This peak spring period is often mistakenly referred to as the "Canadian Seal Hunt", when in fact seal hunting also happens throughout the year all over the Canadian Arctic.

In 2003, the three-year harp seal quota granted by Fisheries and Oceans Canada was increased to a maximum of 975,000 animals per three years, with a maximum of 350,000 animals in any two consecutive years. In 2006, 325,000 harp seals, as well as 10,000 hooded seals and 10,400 grey seals were killed. An additional 10,000 animals were allocated for hunting by aboriginal peoples. As of 2012, the population in Canada of the Northwest Atlantic harp seals is approximately 7.3 million animals, over three times what it was in the 1970s.

Although around 70% of Canadian seals killed are taken on the Front, private monitors focus on the St. Lawrence hunt, because of its more convenient location. The 2006 St. Lawrence leg of the hunt was officially closed on Apr. 3, 2006; sealers had already exceeded the quota by 1,000 animals. On March 26, 2007, the Newfoundland and Labrador government launched a seal hunt website to counter "misinformation about the sealing industry that is [published] by international animal rights organizations".

Warm winters in the Gulf of Saint Lawrence have led to thinner and more unstable ice there. In 2007, Canada's federal fisheries ministry reported that while the pups are born on the ice as usual, the ice floes have started to break up before the pups learn to swim, causing the pups to drown. Canada reduced the 2007 quota by 20%, because overflights showed large numbers of seal pups were lost to thin and melting ice. In southern Labrador and off Newfoundland's northeast coast, however, there was extra heavy ice in 2007, and the coast guard estimated as many as 100 vessels were trapped in ice simultaneously.

The 2010 hunt was cut short because demand for seal pelts was down. Only one local pelt buyer, NuTan Furs, offered to purchase pelts; and it committed to purchase less than 15,000 pelts. Pelt prices were about C$21/pelt in 2010, which is about twice the 2009 price and about 64% of the 2007 price. The reduced demand is attributable mainly to the 2009 ban on imports of seal products into the European Union.

The 2010 winter was unusually warm, with little ice forming in the Gulf of St. Lawrence in February and March, when harp seals give birth to their pups on ice floes. Around the Gulf, harp seals arrived in late winter to give birth on near-shore ice and even on beaches rather than on their usual whelping grounds: sturdy sea ice. Also, seal pups born elsewhere began floating to shore on small, shrinking pieces of ice. Many others stayed too far north, out of reach of all but the most determined hunters. Environment Canada, the weather forecasting agency, reported the ice was at the lowest level on record.

The Fisheries Act established "Seal Protection Regulations" in the mid-1960s. The regulations were combined with other Canadian marine mammals regulations in 1993, to form the "Marine Mammal Regulations". In addition to describing the use of the rifle and hakapik, the regulations state every person "who fishes for seals for personal or commercial use shall land the pelt or the carcass of the seal." The commercial hunting of infant harp seals (whitecoats) and infant hooded seals (bluebacks) was banned in 1987 under pressure from animal rights groups. Now, seals may only be killed once they have started molting (from 12 to 15 days of age), as this coincides with the time when they are abandoned by their mothers.

Export

Canada's biggest market for seal pelts is Norway. Carino Limited is one of Newfoundland's largest seal pelt producers. Carino (CAnada–RIeber–NOrway) is marketing its seal pelts mainly through its parent company, GC Rieber Skinn, Bergen, Norway. Canada sold pelts to eleven countries in 2004. The next largest were Germany, Greenland, and China/Hong Kong. Other importers were Finland, Denmark, France, Greece, South Korea, and Russia. Asia remains the principal market for seal meat exports. One of Canada's market access priorities for 2002 was to "continue to press Korean authorities to obtain the necessary approvals for the sale of seal meat for human consumption in Korea." Canadian and Korean officials agreed in 2003 on specific Korean import requirements for seal meat. For 2004, only Taiwan and South Korea purchased seal meat from Canada.

Canadian seal product exports reached C$18 million in 2006. Of this, C$5.4 million went to the EU. In 2009, the European Union banned all seal imports, shrinking the market. Where pelts once sold for more than $100, they now fetch $8 to $15 each.

Greenland

Although the government of Greenland states that approximately 170,000 seals are killed in Greenland annually, the Government of Canada estimates the kill to be 20,000 to 25,000. In January 2006, the government of Greenland banned imports of Canadian seal skins, citing fears Canadian seals are brutally beaten to death. The boycott may be an effort to distance Greenland's own seal hunt from Canada's, and spare themselves negative press in the process. The ban was rescinded in May 2006, with the Greenland Home Rule Government noting the seal hunt in Canada has sensible regulations on hunting methods, drawn up in close cooperation with biologists, veterinarians, weapons experts and seal hunters. It further noted seal-hunting in Canada is subject to strict and extensive control measures, to ensure the use of effective and humane killing methods.

In Greenland, seal hunting is conducted with rifles – the seals are shot in the head from a small open boat while they sit on an ice floe. The boat rushes up to the seal and hooks the carcass out of the water (where it falls within a few seconds) before it sinks. The economy of certain rural Greenlandic villages, such as Aappilattoq, are highly dependent upon such seal hunting.

Namibia

Year	Annual Quota	Catch
Before 1990	17,000 pups	
1998–2000	30,000 pups	
2001–2003	60,000 pups	
2004–2006	60,000 pups, 7,000 bulls	
2007	80,000 pups, 6,000 bulls	23,000 seals
2008	80,000 pups, 6,000 bulls	
2009	85,000 pups, 7,000 bulls	
2010	85,000 pups, 7,000 bulls	

Namibia is the only country in the Southern Hemisphere culling seals. Although the protection and the sustainable use of natural resources is part of Namibia's constitution, it claims to conduct the second largest seal harvest in the world, mainly because of the huge amount of fish seals are estimated to consume. While a government-initiated study found seal colonies consume more fish than the entire fishing industry can catch, animal protection society Seal Alert South Africa estimated less than 0.3% losses to commercial fisheries.

Harvesting is done from July to November on two places, Cape Cross and Atlas Bay, and in the past at Wolf Bay. These two colonies together account for 75% of the Cape fur seal population of the country.

Cape Cross is a tourism resort and the largest Cape fur seal colony in Namibia. The Department of Tourism has stated that "Cape Cross Seal Reserve was established to protect the largest breeding colony of Cape fur seals in the world". In season, the resort is closed and sealed off during the culling in the early morning hours, journalists are not allowed to enter. Namibia's SPCA is allowed to observe the culling from 2010 onwards.

Namibia's Ministry of Fisheries announced a three-year rolling quota for the seal harvest, although different quotas per year are sometimes reported. The latest quota announced was in 2009, valid until 2011. The quotas are usually not filled by the concession holders.

In 2009, an unusual bid to end seal culling in Namibia was attempted when Seal Alert tried to raise money to purchase the only buyer of Namibian seals, Australian-based Hatem Yavuz, lock, stock, and barrel for US$14.2 million. The project did not materialise. Also, the Government of Namibia offered the International Fund for Animal Welfare (IFAW) an opportunity to buy out the two sealers in Namibia to finally end the culling. The offer was rejected.

In 2011, South African activists launched a boycott of Namibian tourism and Namibian products in response to the cull.

Norway

Hakapiks displayed on the wall of a gun shop in Tromsø, Norway.

The Norwegian sealing vessel Havsel in the West Ice in 2017.

The Norwegian sealing season runs from January to September. The hunt involves "seal catching" by seagoing sealing boats on the Arctic ice shelf, and "seal hunting" on the coast and islands of mainland Norway. The latter is carried out by small groups of licensed hunters shooting seals from land and using small boats to retrieve the catch.

In 2005, Norway allowed foreign nationals to take part in the hunt. In 2006, 17,037 seals (including 13,390 harp and 3,647 hooded seals) were harvested. In 2007, the Norwegian Ministry of Fisheries and Coastal Affairs stated up to 13.5 million Norwegian krone (about US$2.6 million) would be given in funding to vessels in the 2007 Norwegian seal hunt.

All Norwegian sealing vessels are required to carry a qualified veterinary inspector on board. Norwegian sealers are required to pass a shooting test each year before the season starts, using the same weapon and ammunition as they would on the ice. Likewise, they have to pass a hakapik test.

Adult seals more than one year old must be shot in the head with expanding bullets, and cannot be clubbed to death. The hakapik shall be used to ensure the animal is dead. This is done by crushing the skull of the shot adult seal with the short end of the hakapik, before the long spike is thrust deep into the animal's brain. The seal is then bled by making an incision from its jaw to the end of its sternum. The killing and bleeding must be done on the ice, and live animals may never be brought on board the ship. Young seals may be killed using just the hakapik, but only in the aforementioned manner, i.e., they need not be shot.

Seals in the water and seals with young may not be killed, and the use of traps, artificial lighting, airplanes, or helicopters is forbidden.

The hakapik may only be used by certified seal-catchers (fangstmenn) operating in the pack ice of the Arctic Ocean and not by coastal seal hunters. All coastal seal hunters must be pre-approved by the Norwegian Directorate of Fisheries and have to pass a large game hunting test.

In 2007, the European Food Safety Agency confirmed the animals are put to death faster and more humanely in the Norwegian sealing than in large game hunting on land.

Export

In Norway in 2004, only Rieber worked with sealskin and seal oil. In 2001, the biggest producer of raw seal oil was Canada (two percent of the raw oil was processed and sold in Canadian health stores). Rieber had the majority of all distribution of raw seal oil in the world market, but there was no demand for seal oil. From 1995 to 2005, Rieber annually received between 2 and 3 million Norwegian krone in subsidy. A 2003–2004 parliamentary report says CG Rieber Skinn is the only company in the world that delivers skin from bluebacks. Most of the skins processed by Rieber have been imported from abroad, mainly from Canada. Only a small portion is from the Norwegian hunt. Of the processed skin, five percent is sold in Norway; the rest is exported to the Russian and Asian markets.

Fortuna Oils AS (established in 2004) is a 100% owned subsidiary of GC Rieber. They get the majority of their raw oil imported from Canada. They also have access to raw oil from the Norwegian hunt.

Russia

The Russian seal hunt has not been well monitored since the breakup of the Soviet Union. The quota in 1998 was 35,000 animals. Reportedly, many whitecoat pups are not properly killed and are transported, while injured, to processing areas. In January 2000, a bill to ban seal hunting was passed by the Russian parliament by 273 votes to 1, but was vetoed by President Vladimir Putin.

On September 21, 2007 in Arkhangelsk, the Norwegian GC Rieber Skinn AS proposed a joint Russian–Norwegian seal hunting project. The campaign was carried out from one hunt boat supplied by GS Rieber Skinn AS in 2007, lasted two weeks, and brought in 40 000 roubles per Russian hunter. GS Rieber skinn AS declared a plan to order 20 boats and donate them to the Pomor. CG Rieber Skinn AS, in 2007, established a daughter company in Arkhangelsk, called GC Rieber Skinn Pomor'e Lic. (GC Rieber Skinn Pomorje).

The Norwegian company Polardrift AS, in 2007, had plans to establish a company in Russia, and operate under Russian flag, in close cooperation with GC Rieber Skinn Pomor'e.

Plans for the 2008 season included both helicopter-based hunts, mainly to take whitecoats, and boat-based hunts, mainly targeting beaters.

On March 18, 2009, Russia's Minister of Natural Resources and Ecology, Yuriy Trutnev, announced a complete ban on the hunting of harp seals younger than one year of age in the White Sea.

Sealing Debate

Cruelty to Animals

According to a 2002 peer-reviewed study done by five Canadian veterinarians and funded by the Canadian Veterinary Medical Association (CVMA), "the large majority of seals taken during this hunt (at best, 98% in work reported here) are killed in an acceptably humane manner." These veterinarians found, "During the 2001 season in the Gulf, three (1.9%) of 158 seals brought on board of the sealing vessels and directly observed by Daoust had not been killed, and in one (0.86%) of 116 interactions between seals and sealers observed on videotapes by Daoust and Crook, the seal also did not appear to have been killed before being hooked and brought on board." They thus concluded, "This small proportion of animals that are not killed efficiently justifies continued attention to this industry's activities, preferably by members of the veterinary profession, who are best equipped to assess the humaneness of the killing methods."

In observing four videos taken during the 2001 seal hunt in the Gulf of St. Lawrence, the authors of this report state, "A large proportion (87%) of the sealers recorded on the four videotapes failed to palpate the skull or check the corneal reflex before proceeding to hook or bleed the seal or go to another seal."

The Royal Commission on Seals and the Sealing Industry in Canada, also known as the Malouf Commission, concluded in a 1986 report, "Judged by the criteria of rapidity of unconsciousness and particularly the absence of preslaughter stress, the clubbing of seal pups is, when properly performed, at least as humane as, and often more humane than, the killing methods used in commercial slaughterhouses, which are accepted by a majority of the public."

According to the (DFO), "The Marine Mammal Regulations stipulate that seals must be harvested quickly using only high-powered rifles, shotguns firing slugs, clubs or hakapiks.

However, the International Fund for Animal Welfare conducted a study that disputed these findings. This report concludes the Canadian commercial seal hunt results in considerable and unacceptable suffering.

The veterinarians examined 76 seal carcasses and found that in 17% of the cases, there were no detectable lesions of the skull, leading them to conclude the clubbing likely did not result in loss of consciousness. In 25% of the remaining cases, the carcasses had minimal to moderate skull fractures, indicative of a "decreased level of consciousness", but probably not unconsciousness. The remaining 58% of the carcasses examined showed extensive skull fractures.

This veterinary study included examination of video footage of 179 seals hunted in 1998, 1999, and 2000. In these videos, 96 seals were shot, 56 were shot and then clubbed or gaffed, 19 were clubbed or gaffed, and 8 were killed by unknown means. In 79% of these cases, sealers did not check the corneal reflex to ensure that the seals were dead prior to hooking or skinning them. In only 6% of these cases, seals were bled immediately, where struck. The average time from initial strike to bleeding was 66 seconds.

In 2005, IFAW published a comparison of the CVMA-funded study and its own study. In this critique, David M. Lavigne, Science Advisor to IFAW, writes, "The Burdon et al. evidence cited above addresses the question of whether seals were likely conscious or unconscious at the time they were skinned, using post-mortem examination of skulls. In marked contrast, the figure cited from Daoust et al.'s report represents the number of seals clubbed or shot that were brought on board sealing vessels while still conscious. That number ignores any and all animal suffering that occurs between the time animals are clubbed or shot until they eventually reach a sealing vessel, usually on the end of a hook or gaff." Another difference between these reports is "Daoust et al.'s direct observations were made under very different conditions than those provided by Burdon et al. Unlike Burdon et al.'s observations, they were made directly from sealing vessels so that the sealers were unavoidably aware that observers were present. As Daoust et al. admit, the presence of an observer on a sealing vessel "may have incited sealers to hit the seals skulls more vigorously". Of course, the presence of an observer also has the potential to modify other sealing practices, including checking for a corneal reflex and bleeding animals immediately after clubbing."

In 2005, the World Wildlife Fund (WWF) commissioned the Independent Veterinarians Working Group Report. With reference to video evidence, the report states: "Perception of the seal hunt seems to be based largely on emotion, and on visual images that are often difficult even for experienced observers to interpret with certainty. While a hakapik strike on the skull of a seal appears brutal, it is humane if it achieves rapid, irreversible loss of consciousness leading to death."

Ecological Feasibility

In 2013, the Canadian Department of Fisheries and Oceans conducted a population survey. The resulting estimate of the harp seal population was 7.3 million animals, over three times what it was in the 1970s. In 2004, the population estimate was similar: 5.9 million (95% CI 4.6 million to 7.2 million).

Prior to the arrival of European settlers, a much larger population of harp seals migrated to the waters off Newfoundland and Labrador. Settlers began exploiting the population, with kills peaking in the middle of the 1800s. In the first half of the 1840s, 546,000 seals were killed annually. This led to a population decline that adversely affected the industry.

In the 1950s and 1960s an average of over 291,000 seal pups were killed each year. This led to a population decline to less than 2 million seals. Conservationists became alarmed and demanded controls on kill rates. Thus in 1971, Canada instituted a quota system. In the years from 1971 to 1982, an average of 165,627 seals were killed.

In 1983, the European Union banned the import of whitecoat harp seal pup pelts (pelts from pups less than about two weeks of age, when the pups molt). As a result, the market for pelts dropped. The kill rates thus declined in subsequent years to an average of about 52,000 seals from 1983 to 1995. During this time, the harp seal population increased.

After the European Union's ban on whitecoat pelt imports, the Canadian government and sealing industry developed markets for the pelts of beaters. In 1996, the kill rates again increased to over 200,000 each year, except in the year 2000. In 2002 and 2004 to 2006, over 300,000 seal pups were killed each year.

As a result of population concerns, Norway's seal hunt is now controlled by quotas based on recommendations from International Council for the Exploration of the Sea (ICES), However, sealing in Norway has declined in recent years, and the quotas have not been reached.

In addition to hunting pressures on the population of harp seals, as ice seals that are dependent on solid sea ice for whelping, the harp seal population is affected by global climate change. The lack of sea ice in recent years has resulted in the drowning deaths of tens of thousands of newborn harp seal pups.

Objections to Fur

Animal welfare advocates and organizations, such as PETA, object to the use of real fur when many synthetic "faux fur" alternatives are available.

Economic Impact

According to Canadian authorities, the value of the 2004 seal harvest was C$16.5 million, which significantly contributes to seal manufacturing companies, and for several thousand fishermen and First Nations peoples. For some sealers, they claim, proceeds from the hunt make up a third of their annual income. Critics, however, say this represents only a tiny fraction of the C$600-million Newfoundland fishing industry. Sealing opponents also say $16.5 million is insignificant, compared to the funding required to regulate and subsidize the hunt. For 1995 and 1996, there are confirmed reports Fisheries and Oceans Canada encouraged maximum utilization of harvested seals through a $0.20 per pound meat subsidy. The level of subsidy totaled $650,000 in 1997, $440,000 in 1998 and $250,000 in 1999. There were no meat subsidies in 2000. Some critics, such as the McCartneys, have suggested promoting that area as an ecotourism site would be far more lucrative than the annual harvest.

As a Culling Method

In March 2005, Greenpeace asked the DFO to "dispel the myth that seals are hampering the recovery of cod stocks." In doing so, they implied the seal hunt is, at least in part, a cull designed to increase cod stocks. Cod fishing has traditionally been a key part of the Atlantic fishery, and an important part of the economy of Newfoundland and Labrador. Fisheries and Oceans Canada responded there is no connection between the annual seal harvest and the cod fishery, and that the seal hunt is "established on sound conservation principles."

Protests

Many animal protection groups encourage people to petition against the harvest. Respect for Animals and Humane Society International believe the hunt will be ended only by the financial pressure of a boycott of Canadian seafood. In 2005, the Humane Society of the United States (HSUS) called for such a boycott in the United States.

Protesters frequently use images of whitecoats, despite Canada's ban on the commercial hunting of suckling pups. The HSUS explains this by saying images of the legally hunted ragged jackets are nearly indistinguishable from those of whitecoats. Also, they state, according to official DFO kill reports, 97% percent of the estimated million harp seals killed in the last four years have been under three months old, and the majority of these are less than one month old.

Harp seal pup.

On March 26, 2006, seven antisealing activists were arrested in the Gulf of St. Lawrence for violating the terms of their observer permits. By law, observers must maintain a ten-meter distance between themselves and the sealers. Five of the protesters were later acquitted. In the same month, as part of a counterprotest, Newfoundland and Labrador Premier Danny Williams encouraged people in the province to boycott Costco after the retailer decided to stop carrying seal oil capsules. Costco stated politics played no role in their decision to remove the capsules, and on April 4 that year, they were again being sold in Costco stores.

In 2009, the European Union passed a law banning the promotion of imported seal products. The law was approved by the Council of the European Union without debate on July 27, 2009. Denmark, Romania, and Austria abstained. The Canadian government responded to the move by stating that it will take the European Union to the World Trade Organization if the ban does not exempt Canada. Canadian Inuit from the territory of Nunavut have opposed the ban and lobbied European Parliament members against it. Canadian seal hunting issues had been spotlighted in the months leading up to the 2010 Winter Olympics which were held in Vancouver.

Inuit Impact

An important distinction between the southern Canadian seal hunt and the Inuit Canadian hunt is that Canadian Inuit typically hunt ringed seals, while the southern Canadian hunt targets the pelt of the harp seal. Greenlandic Inuit hunt and eat both ringed and harp seals. Unfortunately, protests surrounding the southern seal hunt in Canada have historically damaged the financial well-being of both Canadian and Greenlandic Inuit more than it has the hunting of harp seals. Because of the strongly emotional nature of the protest, profits surrounding the Inuit hunting of ringed seal fall with any ban on any type of seal product, regardless of exemption. For example, in 1963, prices for high-quality pelts of ringed seals could reach above $20.00 each; these same pelts in 1967, the year of the first major protests, would only sell for $2.50 each, which threatened the survival of Inuit in many areas. Several protests in other years have resulted in similar decreases in price.

INTERNATIONAL UNION FOR CONSERVATION OF NATURE

International Union for Conservation of Nature (IUCN), formerly called World Conservation Union is the network of environmental organizations founded as the International Union for the Protection of Nature in October 1948 in Fontainebleau, France, to promote nature conservation and the ecologically sustainable use of natural resources. It changed its name to the International Union for Conservation of Nature and Natural Resources (IUCN) in 1956 and was also known as the World Conservation Union (IUCN) from 1990 to 2008. The IUCN is the world's oldest global environmental organization. Its headquarters are in Gland, Switz.

Through its member organizations, the IUCN supports and participates in environmental scientific research; promotes and helps implement national conservation legislation, policies, and practices; and operates or manages thousands of field projects worldwide. The IUCN's activities are organized into several theme-based programs ranging from business and biodiversity to forest preservation to water and wetlands conservation. In addition, a smaller number of special initiatives draw upon the work of different programs to address specific issues, such as climate change, conservation, and poverty reduction. The volunteer work of more than 10,000 scientists and other experts is coordinated through special commissions on education and communication; environmental, economic, and social policy; environmental law; ecosystem management; species survival; and protected areas. All of the IUCN's work is guided by a global program, which is adopted by member organizations every four years at the IUCN World Conservation Congress.

The IUCN maintains the IUCN Red List of Threatened Species, a comprehensive assessment of the current risk of extinction of thousands of plant and animal species. The organization also publishes or coauthors hundreds of books, reports, and other documents each year. The IUCN has been granted observer status at the United Nations General Assembly.

The IUCN's membership includes more than 1,000 governmental and nongovernmental organizations from more than 140 countries. It is governed by a democratically elected council, which is chosen by member organizations at each World Conservation Congress. The IUCN's funding comes from a number of governments, agencies, foundations, member organizations, and corporations.

MARINE MAMMAL PROTECTION ACT

The Marine Mammal Protection Act (MMPA) was the first act of the United States Congress to call specifically for an ecosystem approach to wildlife management.

MMPA was signed into law on October 21, 1972 by President Richard Nixon and took effect 60 days later on December 21, 1972. It prohibits the "taking" of marine mammals, and enacts a moratorium on the import, export, and sale of any marine mammal, along with any marine mammal part or product within the United States. The Act defines "take" as "the act of hunting, killing, capture, and harassment of any marine mammal; or, the attempt at such." The MMPA defines harassment as "any act of pursuit, torment or annoyance which has the potential to either: a. injure a marine mammal in the wild, or b. disturb a marine mammal by causing disruption of behavioral patterns, which includes, but is not limited to, migration, breathing, nursing, breeding, feeding, or sheltering." The MMPA provides for enforcement of its prohibitions, and for the issuance of regulations to implement its legislative goals.

Authority to manage the MMPA was divided between the Secretary of the Interior through the U.S. Fish and Wildlife Service (Service), and the Secretary of Commerce, which is delegated to the National Oceanic and Atmospheric Administration (NOAA). Subsequently, a third Federal agency, the Marine Mammal Commission (MMC), was established to review existing policies and make recommendations to the Service and the NOAA better implement the MMPA. Coordination between these three Federal agencies is necessary in order to provide the best management practices for marine mammals.

Under the MMPA, the Service is responsible for ensuring the protection of sea otters and marine otters, walruses, polar bears, three species of manatees, and dugongs. NOAA was given responsibility to conserve and manage pinnipeds including seals and sea lions and cetaceans such as whales and dolphins.

Marine Mammal Permits and International Coordination

The MMPA prohibits the take and exploitation of any marine mammal without appropriate authorization, which may only be given by the Service. Permits may be issued for scientific research, public display, and the importation/exportation of marine mammal parts and products upon determination by the Service that the issuance is consistent with the MMPA's regulations. The two types of permits issued by the National Marine Fisheries Service's Office of Protected Resources are incidental and directed. Incidental permits, which allow for some unintentional taking of small numbers of marine mammal, are granted to U.S. citizens who engage in a specified activity other than commercial fishing in a specified geographic area. Directed permits are required for any proposed marine mammal scientific research activity that involves taking marine mammals.

Applications for such permits are reviewed and issued the Service's Division of Management Authority, through the International Affairs office. This office also houses the Division of International Conservation, which is directly responsible for coordinating international activities for marine mammal species found in both U.S. and International waters, or are absent from U.S. waters. Marine mammal species inhabiting both U.S. and International waters include the West Indian manatee, sea otter, polar bear, and Pacific walrus. Species not present in U.S. waters include the West African and Amazonian manatee, dugong, Atlantic walrus, and marine otter.

Marine Mammal Conservation in the Field

In efforts to conserve and manage marine mammal species, the Service has appointed field staff dedicated to working with partners to conduct population censuses, assess population health, develop and implement conservation plans, promulgate regulations, and create cooperative relationships internationally.

Various Marine Mammal Management offices are located on either coast. The Service's Marine Mammal Management office in Anchorage, Alaska is responsible for the management and conservation of polar bears, Pacific walruses, and northern sea otters in Alaska. Northern sea otters present in Washington State are managed by the Western Washington Field Office, while southern sea otters residing in California are managed by the Ventura Field Office. West Indian manatee populations extend from Texas to Rhode Island, and are also present in the Caribbean Sea; however, this species is most prevalent near Florida (the Florida subspecies) and Puerto Rico (the Antillean subspecies). The Service's Jacksonville Field Office manages the Florida manatee, while the Boqueron Field Office manages the Antillean manatee.

The polar bear, southern sea otter, marine otter, all three species of manatees, and the dugong are also concurrently listed under the Endangered Species Act (ESA).

Amendments

Amendments enacted in 1981 established conditions for permits to be granted to take marine mammals "incidentally" in the course of commercial fishing. In addition, the amendments provided additional conditions and procedures for transferring management authority to the States, and authorized appropriations through FY 1984.

- Policies created in 1982:

  ◦ Some marine mammal species or stocks may be in danger of extinction or depletion as a result of human activities.

  ◦ These species or stocks must not be permitted to fall below their optimum sustainable population level (depleted).

  ◦ Measures should be taken to replenish these species or stocks.

  ◦ There is inadequate knowledge of the ecology and population dynamics.

  ◦ Marine mammals have proven to be resources of great international significance.

The 1984 amendments established conditions to be satisfied as a basis for importing fish and fish products from nations engaged in harvesting yellowfin tuna with purse seines and other commercial fishing technology, as well as authorized appropriations for agency activities through FY 1988.

- Amended in 1988:

  ◦ The establishment of conditions and procedures for the Secretaries of Commerce and Interior to review the status of populations to determine if they should be listed as "depleted" (below optimal, sustainable population numbers or listed as threatened or endangered).

  ◦ The preparation of conservation plans for any species listed as depleted, including a requirement that such plans be modeled after recovery plans developed pursuant to the Endangered Species Act.

  ◦ The listing of conditions under which permits may be issued to take marine mammals for the protection and welfare of the animals, including importation, public display, scientific research, and enhancing the survival or recovery of a species.

○   A reward system under which the Secretary of the Treasury can pay up to $2500 to individuals providing information leading to convictions for violations of the Act.

- Amended in 1995:

    ○   Certain exceptions to the take prohibitions, such as for Alaska Native subsistence and permits and authorizations for scientific research.

    ○   A program to authorize and control the taking of marine mammals incidental to commercial fishing operations.

    ○   Preparation of stock assessments for all marine mammal stocks in waters under U.S. jurisdiction.

    ○   Studies of pinniped-fishery interactions.

Findings

Congress found that: all species and population stocks of marine mammals are, or may be, in danger of extinction or depletion due to human activities; these mammals should not be permitted to diminish below their optimum sustainable population; measures should be taken immediately to replenish any of these mammals that have diminished below that level, and efforts should be made to protect essential habitats; there is inadequate knowledge of the ecology and population dynamics of these mammals; negotiations should be undertaken immediately to encourage international arrangements for research and conservation of these mammals. Congress declared that marine mammals are resources of great international significance (aesthetic, recreational and economic), and should be protected and encouraged to develop to the greatest extent feasible commensurate with sound policies of resource management. The primary management objective should be to maintain the health and stability of the marine ecosystem. The goal is to obtain an optimum sustainable population within the carrying capacity of the habitat.

MARINE PROTECTED AREA

Marine protected areas (MPA) are protected areas of seas, oceans, estuaries or in the US, the Great Lakes. These marine areas can come in many forms ranging from wildlife refuges to research facilities. MPAs restrict human activity for a conservation purpose, typically to protect natural or cultural resources. Such marine resources are protected by local, state, territorial, native, regional, national, or international authorities and differ substantially among and between nations. This variation includes different limitations on development, fishing practices, fishing seasons and catch limits, moorings and bans on removing or disrupting marine life. In some situations (such as with the Phoenix Islands Protected Area), MPAs also provide revenue for countries, potentially equal to the income that they would have if they were to grant companies permissions to fish.

On 28 October 2016 in Hobart, Australia, the Convention for the Conservation of Antarctic Marine Living Resources agreed to establish the first Antarctic and largest marine protected area in the world encompassing 1.55 million km2 (600,000 sq mi) in the Ross Sea. Other large MPAs are in the Indian, Pacific, and Atlantic Oceans, in certain exclusive economic zones of Australia and overseas territories of France, the United Kingdom and the United States, with major (990,000 square kilometres (380,000 sq mi) or larger) new or expanded MPAs by these nations since 2012—such as Natural Park of the Coral Sea, Pacific Remote Islands Marine National Monument, Coral Sea Commonwealth Marine Reserve and South Georgia and the South Sandwich Islands Marine Protected Area. When counted with MPAs of all sizes from many other countries, as of August 2016 there are more than 13,650 MPAs, encompassing 2.07% of the world's oceans, with half of that area – encompassing 1.03% of the world's oceans – receiving complete "no-take" designation.

Classifications

The Chagos Archipelago was declared the world's largest marine reserve in April 2010 with an area of 250,000 square miles until March 2015 when It was declared illegal by the Permanent Court of Arbitration.

Several types of compliant MPA can be distinguished:

- A totally marine area with no significant terrestrial parts.

- An area containing both marine and terrestrial components, which can vary between two extremes; those that are predominantly maritime with little land (for example, an atoll would have a tiny island with a significant maritime population surrounding it), or that is mostly terrestrial.

- Marine ecosystems that contain land and intertidal components only. For example, a mangrove forest would contain no open sea or ocean marine environment, but its river-like marine ecosystem nevertheless complies with the definition.

IUCN offered seven categories of protected area, based on management objectives and four broad governance types.

Categories	IUCN Protected Area Management Categories
Ia	Strict nature reserve • A marine reserve usually connotes "maximum protection", where all resource removals are strictly prohibited. In countries such as Kenya and Belize, marine reserves allow for low-risk removals to sustain local communities.
Ib	Wilderness area

II	National park • Marine parks emphasize the protection of ecosystems but allow light human use. A marine park may prohibit fishing or extraction of resources, but allow recreation. Some marine parks, such as those in Tanzania, are zoned and allow activities such as fishing only in low risk areas.
III	Natural monuments or features • Established to protect historical sites such as shipwrecks and cultural sites such as aboriginal fishing grounds.
IV	Habitat/species management area • Established to protect a certain species, to benefit fisheries, rare habitat, as spawning/nursing grounds for fish, or to protect entire ecosystems.
V	Protected seascape • Limited active management, as with protected landscapes.
VI	Sustainable use of natural resources

Related protected area categories include the following;

- World Heritage Site (WHS) – an area exhibiting extensive natural or cultural history. Maritime areas are poorly represented, however, with only 46 out of over 800 sites.

- Man and the Biosphere – UNESCO program that promotes "a balanced relationship between humans and the biosphere". Under article 4, biosphere reserves must "encompass a mosaic of ecological systems", and thus combine terrestrial, coastal, or marine ecosystems. In structure they are similar to Multiple-use MPAs, with a core area ringed by different degrees of protection.

- Ramsar site – must meet certain criteria for the definition of "Wetland" to become part of a global system. These sites do not necessarily receive protection, but are indexed by importance for later recommendation to an agency that could designate it a protected area.

While "area" refers to a single contiguous location, terms such as "network", "system", and "region" that group MPAs are not always consistently employed."System" is more often used to refer to an individual MPA, whereas "region" is defined by the World Conservation Monitoring Centre as:

"A collection of individual MPAs operating cooperatively, at various spatial scales and with a range of protection levels that are designed to meet objectives that a single reserve cannot achieve."

At the 2004 Convention on Biological Diversity, the agency agreed to use "network" on a global level, while adopting system for national and regional levels. The network is a mechanism to establish regional and local systems, but carries no authority or mandate, leaving all activity within the "system".

No take zones (NTZs), are areas designated in a number of the world's MPAs, where all forms of exploitation are prohibited and severely limits human activities. These no take zones can cover an entire MPA, or specific portions. For example, the 1,150,000 square kilometres (440,000 sq mi) Papahānaumokuākea Marine National Monument, the world's largest MPA (and largest protected area of any type, land or sea), is a 100% no take zone.

Related terms include; specially protected area (SPA), Special Area of Conservation (SAC), the United Kingdom's marine conservation zones (MCZs), or area of special conservation (ASC) etc. which each provide specific restrictions.

Stressors

Stressors that affect oceans include "the impact of extractive industries, localised pollution, and changes to its chemistry (ocean acidification) resulting from elevated carbon dioxide levels, due to our emissions". MPAs have been cited as the ocean's single greatest hope for increasing the resilience of the marine environment to such stressors. Well-designed and managed MPAs developed with input and support from interested stakeholders can conserve biodiversity and protect and restore fisheries.

Economics

MPAs can help sustain local economies by supporting fisheries and tourism. For example, Apo Island in the Philippines made protected one quarter of their reef, allowing fish to recover, jump-starting their economy. This was shown in the film, Resources at Risk: Philippine Coral Reef. A 2016 report by the Center for Development and Strategy found that programs like the United States National Marine Sanctuary system can develop considerable economic benefits for communities through Public–private partnerships.

Management

Typical MPAs restrict fishing, oil and gas mining and/or tourism. Other restrictions may limit the use of ultrasonic devices like sonar (which may confuse the guidance system of cetaceans), development, construction and the like. Some fishing restrictions include "no-take" zones, which means that no fishing is allowed. Less than 1% of US MPAs are no-take.

Ship transit can also be restricted or banned, either as a preventive measure or to avoid direct disturbance to individual species. The degree to which environmental regulations affect shipping varies according to whether MPAs are located in territorial waters, exclusive economic zones, or the high seas. The law of the sea regulates these limits.

Asinara, Italy is listed by WDPA as both a marine reserve and
a national marine park, and as such could be labelled 'multiple-use'.

Most MPAs have been located in territorial waters, where the appropriate government can enforce them. However, MPAs have been established in exclusive economic zones and in international waters. For example, Italy, France and Monaco in 1999 jointly established a cetacean sanctuary in the Ligurian Sea named the Pelagos Sanctuary for Mediterranean Marine Mammals. This sanctuary includes both national and international waters. Both the CBD and IUCN recommended a variety of management systems for use in a protected area system. They advocated that MPAs be seen as one of many "nodes" in a network of protected areas. The following are the most common management systems:

- Seasonal and temporary management—Activities, most critically fishing, are restricted seasonally or temporarily, e.g., to protect spawning/nursing grounds or to let a rapidly reducing species recover.

- Multiple-use MPAs—These are the most common and arguably the most effective. These areas employ two or more protections. The most important sections get the highest protection, such as a no take zone and are surrounded with areas of lesser protections.

- Community involvement and related approaches—Community-managed MPAs empower local communities to operate partially or completely independent of the governmental jurisdictions they occupy. Empowering communities to manage resources can lower conflict levels and enlist the support of diverse groups that rely on the resource such as subsistence and commercial fishers, scientists, recreation, tourism businesses, youths and others.

Marine Protected Area Networks

Marine Protected Area Networks or MPA networks have been defined as "A group of MPAs that interact with one another ecologically and/or socially form a network".

These networks are intended to connect individuals and MPAs and promote education and cooperation among various administrations and user groups. "MPA networks are, from the perspective of resource users, intended to address both environmental and socio-economic needs, complementary ecological and social goals and designs need greater research and policy support".

Filipino communities connect with one another to share information about MPAs, creating a larger network through the social communities' support. Emerging or established MPA networks can be found in Australia, Belize, the Red Sea, Gulf of Aden and Mexico.

International Efforts

The 17th International Union for Conservation of Nature (IUCN) General Assembly in San Jose, California, the 19th IUCN assembly and the fourth World Parks Congress all proposed to centralise the establishment of protected areas. The World Summit on Sustainable Development in 2002 called for the establishment of marine protected areas consistent with international laws and based on scientific information, including representative networks by 2012.

The Evian agreement, signed by G8 Nations in 2003, agreed to these terms. The Durban Action Plan, developed in 2003, called for regional action and targets to establish a network of protected areas by 2010 within the jurisdiction of regional environmental protocols.It recommended establishing protected areas for 20 to 30% of the world's oceans by the goal date of 2012. The Convention on Biological Diversity considered these recommendations and recommended requiring countries to set up marine parks controlled by a central organization before merging them. The United Nations Framework Convention on Climate Change agreed to the terms laid out by the convention, and in 2004, its member nations committed to the following targets:

- By 2006 complete an area system gap analysis at national and regional levels.

- By 2008 address the less represented marine ecosystems, accounting for those beyond national jurisdiction in accordance.

- By 2009 designate the protected areas identified through the gap analysis.

- By 2012 complete the establishment of a comprehensive and ecologically representative network.

Bunaken Marine Park, Indonesia is officially listed as
both a marine reserve and a national marine park.

"The establishment by 2010 of terrestrial and by 2012 for marine areas of comprehensive, effectively managed, and ecologically representative national and regional systems of protected areas that collectively, inter alia through a global network, contribute to achieving the three objectives of the Convention and the 2010 target to significantly reduce the current late of biodiversity loss at the global, regional, national, and sub-national levels and contribute to poverty reduction and the pursuit of sustainable development."

The UN later endorsed another decision, Decision VII/15, in 2006:

Effective conservation of 10% of each of the world's ecological regions by 2010.

The Antarctic Treaty System

On 7 April 1982, the Convention on the Conservation of Antarctic Marine Living Resources (CAMLR Convention) came into force after discussions began in 1975 between parties of the then-current Antarctic Treaty to limit large-scale exploitation of krill by commercial fisheries. The Convention bound contracting nations to abide by previously agreed upon Antarctic territorial claims and peaceful use of the region while protecting ecosystem integrity south of the Antarctic Convergence and 60 S latitude. In so doing, it also established a commission of the original signatories and acceding parties called the Commission for the Conservation of Antarctic Marine Living Resources (CCAMLR) to advance these aims through protection, scientific study, and rational use, such as harvesting, of those marine resources. Though separate, the Antarctic Treaty and CCAMLR, make up part the broader system of international agreements called the Antarctic Treaty System. Since 1982, the CCAMLR meets annually to implement binding conservations measures like the creation of 'protected areas' at the suggestion of the convention's scientific committee.

In 2009, the CCAMLR created the first 'high-seas' MPA entirely within international waters over the southern shelf of the South Orkney Islands. This area encompasses 94,000 square kilometres (36,000 sq mi) and all fishing activity including transhipment, and dumping or discharge of waste is prohibited with the exception of scientific research endeavors. On 28 October 2016, the CCAMLR, composed of 24 member countries and the European Union at the time, agreed to establish the world's largest marine park encompassing 1.55 million km² (600,000 sq mi) in the Ross Sea after several years of failed negotiations. Establishment of the Ross Sea MPA required unanimity of the commission members and enforcement will begin in December 2017. However, due to a sunset provision inserted into the proposal, the new marine park will only be in force for 35 years.

National Efforts

The marine protected area network is still in its infancy. As of October 2010, approximately 6,800 MPAs had been established, covering 1.17% of global ocean area. Protected areas covered 2.86% of exclusive economic zones (EEZs). MPAs covered 6.3% of territorial seas. Many prohibit the use of harmful fishing techniques yet only 0.01% of the ocean's area is designated as a "no take zone". This coverage is far below the projected goal of 20%-30% Those targets have been questioned mainly due to the cost of managing protected areas and the conflict that protections have generated with human demand for marine goods and services.

Africa

South Africa

A marine protected area of South Africa is an area of coastline or ocean within the exclusive economic zone (EEZ) of the Republic of South Africa that is protected in terms of specific legislation.

There are a total of 45 marine protected areas in the South African EEZ, with a total area of 5% of the waters. The target is to have 10% of the oceanic waters protected by 2020. All but one of the MPAs are in the coastal waters off continental South Africa, and one is off Prince Edward Island in the Southern Ocean.

Greater Caribbean

The Greater Caribbean subdivision encompasses an area of about 5,700,000 square kilometres (2,200,000 sq mi) of ocean and 38 nations. The area includes island countries like the Bahamas and Cuba, and the majority of Central America. The Convention for Protection and Development of the Marine Environment of the Wider Caribbean Region (better known as the Cartagena Convention) was established in 1983. Protocols involving protected areas were ratified in 1990. As of 2008, the region hosted about 500 MPAs. Coral reefs are the best represented.

Two networks are under development, the Mesoamerican Barrier Reef System (a long barrier reef that borders the coast of much of Central America), and the "Islands in the Stream" program (covering the Gulf of Mexico).

Southeast Asia

Southeast Asia is a global epicenter for marine diversity. 12% of its coral reefs are in MPAs. The Philippines have some the world's best coral reefs and protect them to attract international tourism. Most of the Philippines' MPAs are established to secure protection for its coral reef and sea grass habitats. Indonesia has MPAs designed for tourism and relies on tourism as a main source of income.

Philippines

The Philippines host one of the most highly biodiverse regions, with 464 reef-building coral species. Due to overfishing, destructive fishing techniques, and rapid coastal development, these are in rapid decline. The country has established some 600 MPAs. However, the majority are poorly enforced and are highly ineffective. However, some have positively impacted reef health, increased fish biomass, decreased coral bleaching and increased yields in adjacent fisheries. One notable example is the MPA surrounding Apo Island.

Latin America

Latin America has designated one large MPA system. As of 2008, 0.5% of its marine environment was protected, mostly through the use of small, multiple-use MPAs.

South Pacific

The South Pacific network ranges from Belize to Chile. Governments in the region adopted the Lima convention and action plan in 1981. An MPA-specific protocol was ratified in 1989. The permanent commission on the exploitation and conservation on the marine resources of the South Pacific promotes the exchange of studies and information among participants.

The region is currently running one comprehensive cross-national program, the Tropical Eastern Pacific Marine Corridor Network, signed in April 2004. The network covers about 211,000,000 square kilometres (81,000,000 sq mi).

One alternative to imposing MPAs on an indigenous population is through the use of Indigenous Protected Areas, such as those in Australia.

North Pacific

The North Pacific network covers the western coasts of Mexico, Canada, and the U.S. The "Antigua Convention" and an action plan for the north Pacific region were adapted in 2002. Participant nations manage their own national systems. In 2010-2011, the State of California completed hearings and actions via the state Department of Fish and Game to establish new MPAs.

United States and Pacific Island Territories

President Barack Obama signed a proclamation on September 25, 2014, designating the world's largest marine reserve. The proclamation expanded the existing Pacific Remote Islands Marine National Monument, one of the world's most pristine tropical marine environments, to six times its current size, encompassing 490,000 square miles (1,300,000 km²) of protected area around these islands. Expanding the Monument protected the area's unique deep coral reefs and seamounts.

The orientation of the 3 marine sanctuaries of Central California: Cordell Bank, Gulf of the Farallones, and Monterey Bay. Davidson Seamount, part of the Monterey Bay sanctuary, is indicated at bottom-right.

In April 2009, the US established a United States National System of Marine Protected Areas, which strengthens the protection of US ocean, coastal and Great Lakes resources. These large-scale MPAs should balance "the interests of conservationists, fishers, and the public." As of 2009, 225 MPAs participated in the national system. Sites work together toward common national and regional conservation goals and priorities. NOAA's national marine protected areas center maintains a comprehensive inventory of all 1,600+ MPAs within the US exclusive economic zone. Most US MPAs. Allow some type of extractive use. Fewer than 1% of U.S. waters prohibit all extractive activities.

In 1981 Olympic National Park became a marine protected area. The total protected site area is 3,697 square kilometres (1,427 sq mi). 173.2 km² of the area was an MPA. The national system is a mechanism to foster MPA collaboration. Sites that meet pertinent criteria are eligible to join the national system. Four entry criteria govern admission:

- Meets the definition of an MPA as defined in the Framework.

- Has a management plan (can be sitespecific or part of a broader programmatic management plan; must have goals and objectives and call for monitoring or evaluation of those goals and objectives).

- Contributes to at least one priority conservation objective as listed in the Framework.

- Cultural heritage MPAs must also conform to criteria for the National Register for Historic Places."

In 1999, California adopted the Marine Life Protection Act, establishing the first state law requiring a comprehensive, science-based MPA network. The state created the Marine Life Protection Act Initiative. The MLPA Blue Ribbon Task Force and stakeholder and scientific advisory groups ensure that the process uses the science and public participation.

The MLPA Initiative established a plan to create California's statewide MPA network by 2011 in several steps. The Central Coast step was successfully completed in September, 2007. The North Central Coast step was completed in 2010. The South Coast and North Coast steps were expected to go into effect in 2012.

Indian Ocean

In exchange for some of its national debt being written off, the Seychelles designates two new marine protected areas in the Indian Ocean, covering about 210,000 square kilometres (81,000 sq mi). It is the result of a financial deal, brokered in 2016 by The Nature Conservancy.

United Kingdom and British Overseas Territories

There are a number of marine protected areas around the coastline of the United Kingdom, known as Marine Conservation Zones in England, Wales, and Northern Ireland, Marine Protection Areas in Scotland. They are to be found in inshore and offshore waters.

The United Kingdom is also creating marine protected reserves around several British Overseas Territories. The UK is responsible for 6.8 million square kilometres of ocean around the world, larger than all but four other countries.

The Chagos Marine Protected Area in the Indian Ocean was established in 2010 as a "no-take-zone". With a total surface area of 640,000 square kilometres (250,000 sq mi), it was the world's largest contiguous marine reserve. In March 2015, the UK announced the creation of a marine reserve around the Pitcairn Islands in the Southern Pacific Ocean to protect its special biodiversity. The area of 830,000 square kilometres (320,000 sq mi) surpassed the Chagos Marine Protected Area as the world's largest contiguous marine reserve, until the August 2016 expansion of the Papahānaumokuākea Marine National Monument in the United States to 1,510,000 square kilometres (580,000 sq mi).

In January 2016, the UK government announced the intention to create a marine protected area around Ascension Island. The protected area will be 234,291 square kilometres (90,460 sq mi), half of which will be closed to fishing.

Europe

The Natura 2000 ecological MPA network in the European Union included MPAs in the North Atlantic, the Mediterranean Sea and the Baltic Sea. The member states had to define NATURA 2000 areas at sea in their Exclusive Economic Zone.

Two assessments, conducted thirty years apart, of three Mediterranean MPAs, demonstrate that proper protection allows commercially valuable and slow-growing red coral (Corallium rubrum) to produce large colonies in shallow water of less than 50 metres (160 ft). Shallow-water colonies outside these decades-old MPAs are typically very small. The MPAs are Banyuls, Carry-le-Rouet and Scandola, off the island of Corsica.

- Mediterranean Science Commission; proposed the creation of 7 marine protected areas ("peace parcs").

- WWF together with other partners proposed the creation of MedPan (Network of Marine Protected Areas Managers in the Mediterranean) which aims to protect 10% of the surface of the mediterranean by 2020.

Notable Marine Protected Areas

- The Bowie Seamount Marine Protected Area off the coast of British Columbia, Canada.

- The Great Barrier Reef Marine Park in Queensland, Australia.

- The Ligurian Sea Cetacean Sanctuary in the seas of Italy, Monaco and France.

- The Dry Tortugas National Park in the Florida Keys, USA.

- The Papahānaumokuākea Marine National Monument in Hawaii.

- The Phoenix Islands Protected Area, Kiribati.

- The Channel Islands National Marine Sanctuary in California, USA.

- The Chagos Marine Protected Area in the Indian Ocean.

- The Wadden Sea bordering the North Sea in the Netherlands, Germany, and Denmark.

Assessment

Managers and scientists use geographic information systems and remote sensing to map and analyze MPAs. NOAA Coastal Services Center compiled an "Inventory of GIS-Based Decision-Support Tools for MPAs." The report focuses on GIS tools with the highest utility for MPA processes. Remote sensing uses advances in aerial photography image capture, pop-up archival satellite tags, satellite imagery, acoustic data, and radar imagery. Mathematical models that seek to reflect the complexity of the natural setting may assist in planning harvesting strategies and sustaining fishing grounds.

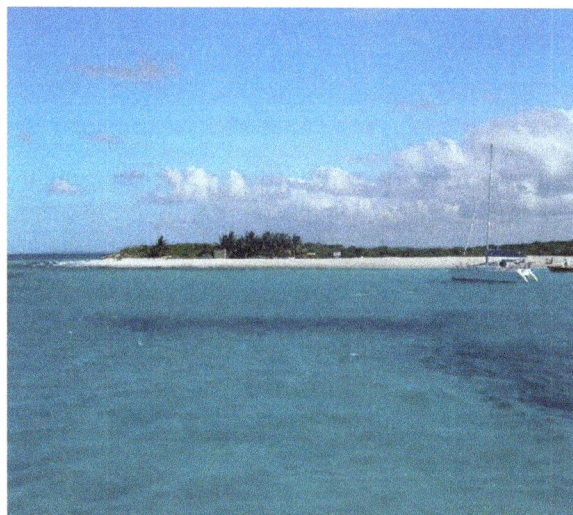

The Prickly Pear Cays are a marine protected area, roughly six miles from
Road Bay, Anguilla, in the Leeward Islands of the Caribbean.

Coral Reefs

Coral reef systems have been in decline worldwide. Causes include overfishing, pollution and ocean acidification. As of 2013 30% of the world's reefs were severely damaged. Approximately 60% will be lost by 2030 without enhanced protection. Marine reserves with "no take zones" are the most effective form of protection. Only about 0.01% of the world's coral reefs are inside effective MPAs.

Fish

MPAs can be an effective tool to maintain fish populations. The general concept is to create overpopulation within the MPA. The fish expand into the surrounding areas to reduce crowding, increasing the population of unprotected areas. This helps support local fisheries in the surrounding area, while maintaining a healthy population within the MPA. Such MPAs are most commonly used for coral reef ecosystems.

One example is at Goat Island Bay in New Zealand, established in 1977. Research gathered at Goat Bay documented the spillover effect. "Spillover and larval export—the drifting of millions of eggs and larvae beyond the reserve—have become central concepts of marine conservation". This positively impacted commercial fishermen in surrounding areas.

Another unexpected result of MPAs is their impact on predatory marine species, which in some conditions can increase in population. When this occurs, prey populations decrease. One study showed that in 21 out of 39 cases, "trophic cascades," caused a decrease in herbivores, which led to an increase in the quantity of plant life. (This occurred in the Malindi Kisite and Watamu Marian National Parks in Kenya; the Leigh Marine Reserve in New Zealand; and Brackett's Landing Conservation Area in the US.

Success Criteria

Both CBD and IUCN have criteria for setting up and maintaining MPA networks, which emphasize 4 factors:

- Adequacy: Ensuring that the sites have the size, shape, and distribution to ensure the success of selected species.

- Representability: Protection for all of the local environment's biological processes

- Resilience: The resistance of the system to natural disaster, such as a tsunami or flood.

- Connectivity: Maintaining population links across nearby MPAs.

Misconceptions

Misconceptions about MPAs include the belief that all MPAs are no-take or no-fishing areas. However, less than 1 percent of US waters are no-take areas. MPA activities can include consumption fishing, diving and other activities.

Another misconception is that most MPAs are federally managed. Instead, MPAs are managed under hundreds of laws and jurisdictions. They can exist in state, commonwealth, territory and tribal waters.

Another misconception is that a federal mandate dedicates a set percentage of ocean to MPAs. Instead the mandate requires an evaluation of current MPAs and creates a public resource on current MPAs.

Criticism

Some existing and proposed MPAs have been criticized by indigenous populations and their supporters, as impinging on land usage rights. For example, the proposed Chagos Protected Area in the Chagos Islands is contested by Chagossians deported from their homeland in 1965 by the British as part of the creation of the British Indian Ocean Territory (BIOT). The UK proposed that the BIOT become a "marine reserve" with the aim of preventing the former inhabitants from returning to their lands and to protect the joint UK/US military base on Diego Garcia Island.

Other critiques include: their cost (higher than that of passive management), conflicts with human development goals, inadequate scope to address factors such as climate change and invasive species.

References

- Brunborg, Linn Anne; Julshamn, Kare; Nortvedt, Ragnar; Frøyland, Livar (2006). "Nutritional composition of blubber and meat of hooded seal and harp seal". Food Chemistry. 96 (4): 524–531. Doi:10.1016/j.foodchem.2005.03.005. Retrieved 2011-03-03

- Biggest-threats-marine-mammals-face-today, animalsandnature: onegreenplanet.org, Retrieved 13 June, 2019

- Ejesiak, Kirt; et al. (8 May 2005), Animal rights vs. Inuit rights, Boston Globe, archived from the original on November 4, 2012, Retrieved 8 May 2010

- International-Union-for-Conservation-of-Nature, topic: britannica.com, Retrieved 14 July, 2019

- Kuempel, Caitlin; Jones, Kendall; Watson, James; Possingham, James (27 May 2017). "Quantifying biases in marine-protected-area placement relative to abatable threats". Conservation Biology. Doi:10.1111/cobi.13340. PMID 31131932

- "CCAMLR to create world›s largest Marine Protected Area". Commission for the Conservation of Antarctic Marine Living Resources. 29 October 2016. Retrieved 29 October 2016

Permissions

Index